Amplifier Circuits

Other Books in this Series

Amplifier Circuits

Rudolf F. Graf

Newnes
Boston   Oxford   Johannesburg   Melbourne   New Delhi   Singapore

Newnes is an imprint of Butterworth-Heinemann

Copyright © 1997 by Butterworth–Heinemann
Copyright © 1992 by Rudolf F. Graf

ℛ A member of the Reed Elsevier group

Library of Congress Cataloging-in-Publication Data
Graf, Rudolf F.
    [Modern amplifier circuit encyclopedia]
    Amplifier circuits / Rudolf F. Graf.
        p.   cm.
    Originally published: The modern amplifier circuit encyclopedia.
    1st ed. Blue Ridge Summit, PA : TAB Books, c1992.
    Includes index.
    ISBN 0-7506-9877-2
    1. Amplifiers (Electronics) I. Title. II. Title: Amplifier
circuit.
TK7871.2.G73   1996
621.381'535—dc20   96-36498
                                    CIP

British Library Cataloguing-in-Publication Data
A catalogue record for this book is available from the British Library.

The publisher offers special discounts on bulk orders of this book.
For information, please contact:
Manager of Special Sales
Butterworth–Heinemann
313 Washington Street
Newton, MA 02158–1626
Tel: 617-928-2500
Fax: 617-928-2620

For information on all Newnes electronics publications available, contact
our World Wide Web home page at: http://www.bh.com/bh

Printed in the United States of America
10 9 8 7 6 5 4 3 2 1

Contents

Introduction

Like the other volumes in this series, this book contains a wealth of ready-to-use circuits that serve the needs of the engineer, technician, student and, of course, the browser. These unique books contain more practical, ready-to-use circuits focused on a specific field of interest, than can be found anywhere in a single volume.

1

Audio Power Amplifiers

The sources of the following circuits are contained in the Sources section, which begins on page 182. The figure number contained with each circuit correlates to the source entry in the Sources section.

HYBRID POWER AMPLIFIER

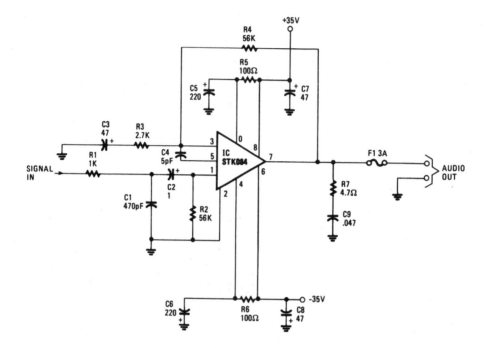

Fig. 1-1

The input is ac coupled to the amplifier through C2, which blocks dc signals that might also be present at the input. The R1/C1 combination forms a low-pass filter, which eliminates unwanted high-frequency signals by bypassing them to ground when they appear at the circuit input, which has an impedance of about 52 Ω. The gain of the amplifier is set at about 26 dB by resistors R3 and R4. The R5/C5/C7 combination on the positive supply and its counterpart R6/C6/C8 on the negative supply provides power-supply decoupling. R7 and C9 together prevent oscillation at the output of the amplifier. From that point, the amplifier's output signal is direct coupled to the speaker through a 3-A fuse, F1. The dc output of the .amplifier at pin 7 is 0 V, so no dc current flows through the speaker. Should there be a catastrophic failure of the output stage, fuse F1, which should be a fast-acting type, prevents dc from flowing through the speaker.

BRIDGE AMPLIFIER

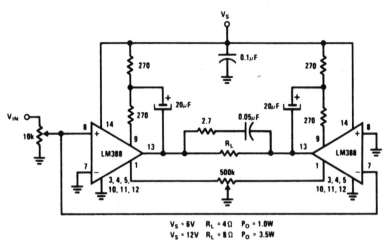

$$V_S = 6V \quad R_L = 4\,\Omega \quad P_O = 1.0W$$
$$V_S = 12V \quad R_L = 8\,\Omega \quad P_O = 3.5W$$

NATIONAL SEMICONDUCTOR

Fig. 1-2

This circuit is for low-voltage applications that require high power outputs. Output power levels of 1.0 W into 4 Ω from 6 V and 3.5 V into 8 Ω from 12 V are typical. Coupling capacitors are not necessary because the output dc levels will be within a few tenths of a volt of each other. Where critical matching is required, the 500-kΩ potentiometer is added and adjusted for zero dc current flow through the load.

NONINVERTING AMPLIFIER USING SINGLE SUPPLY

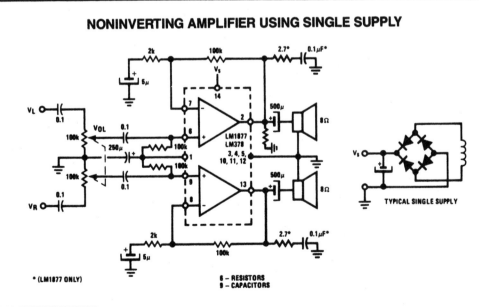

* (LM1877 ONLY)

6 – RESISTORS
9 – CAPACITORS

TYPICAL SINGLE SUPPLY

NATIONAL SEMICONDUCTOR

Fig. 1-3

OUTPUT-STAGE POWER BOOSTER

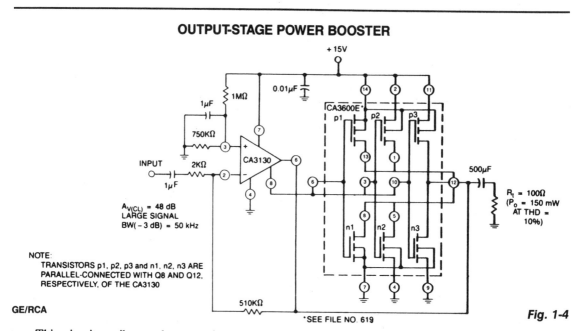

NOTE:
TRANSISTORS p1, p2, p3 and n1, n2, n3 ARE
PARALLEL-CONNECTED WITH Q8 AND Q12,
RESPECTIVELY, OF THE CA3130

$A_{V(CL)}$ = 48 dB
LARGE SIGNAL
BW(–3 dB) = 50 kHz

R_1 = 100Ω
(P_o = 150 mW
AT THD =
10%)

GE/RCA

*SEE FILE NO. 619

Fig. 1-4

This circuit easily supplements the current-sourcing and current-sinking capability of the CA3130 BiMOS op amp. This arrangement boosts the current-handling capability of the CA3130 output stage by about 2.5 times.

PORTABLE AMPLIFIER

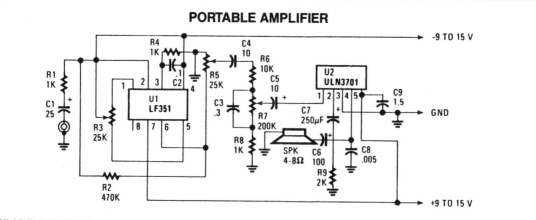

POPULAR ELECTRONICS

Fig. 1-5

U1, an FET op amp needs a bipolar voltage at pins 4 and 7 with a common ground for optimum gain. You can calculate the gain by dividing R2 by R1. Zero-set balance can be had through pins 1 and 5 through R3. Put a voltmeter between pin 6 and ground and adjust R3 for zero voltage. Once you've established that, you can measure the ohmic resistance at each side of R3's center tap and replace the potentiometer with fixed resistors. R6, R7, R8, and C3 form a tone control that will give you added bass boost, if needed.

4

REAR-SPEAKER AMBIENCE (4-CHANNEL) AMPLIFIER

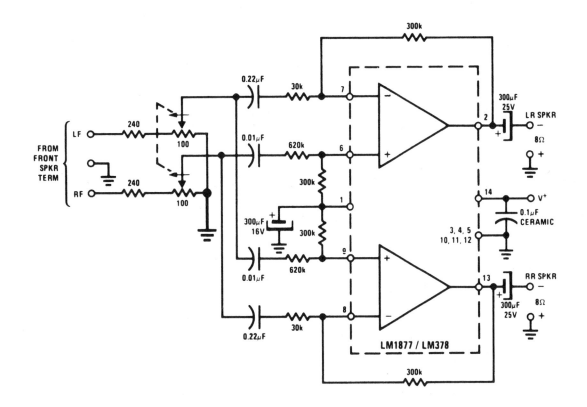

NATIONAL SEMICONDUCTOR CORP.

Fig. 1-6

Rear channel "ambience" can be added to an existing stereo system to extract a difference signal (R − L or L − R) which, when combined with some direct signal (R or L), adds fullness, or "concert hall realism" to the reproduction of recorded music. Very little power is required at the rear channels, hence an LM1877 suffices for most "ambience" applications. The inputs are merely connected to the existing speaker output terminals of a stereo set, and two more speakers are connected to the ambience circuit outputs. The rear speakers should be connected in the opposite phase to those of the front speakers, as indicated by the +/− signs.

WALKMAN AMPLIFIER

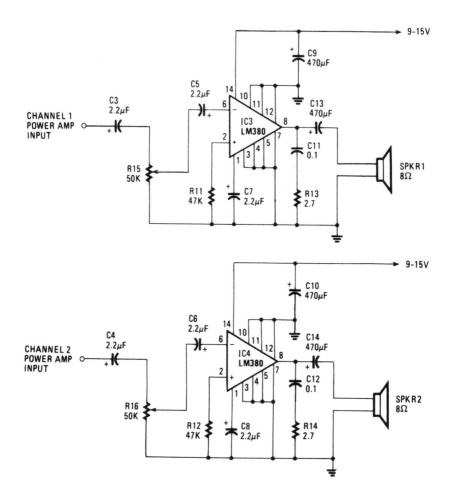

Fig. 1-7

The gain of the low-cost IC is internally fixed so that it is not less than 34 dB (50 times). A unique input stage allows input signals to be referenced to ground. The output is automatically self centering to one half the supply voltage. The output is also short-circuit proof with internal thermal limiting. With a maximum supply of 15 V and an 8-Ω load, the output is near 1.5 W per channel. The input stage is usable

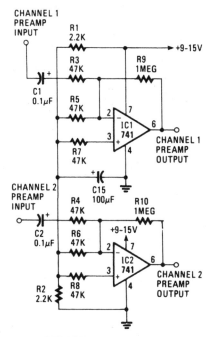

—THE PREAMP. If you wish to amplify low-level signals, such as the output of a turntable, the signal will first have to be fed to the preamp shown here.

with signals from 50 to 500 mV rms. If the amplifier is to be used with a source other than a personal stereo, such as a phonograph or an electric guitar, some type of preamplifier is required. A suitable circuit is shown. In that circuit, two 741 op amps have been configured as input amplifiers. Their input stages referenced to a common point—half the supply voltage. That voltage is derived from a voltage divider made up of R1 and R2, two 2.2-kΩ resistors. The gain of each of the 741's has been fixed at 21 by the input resistors (R9, R10). Input capacitors, C1 and C2, are used to filter out any dc component from the input signal.

SPEAKER AMPLIFIER FOR HAND-HELD TRANSCEIVERS

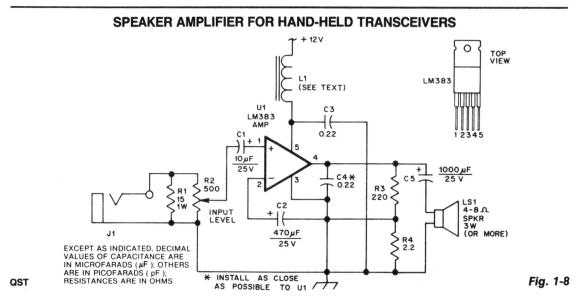

QST

Fig. 1-8

The LM383 is an audio-power amplifier that is capable of producing up to 8 W of audio output. R1 is essentially a load resistor for the hand-held transceiver's audio output. R2 can be composed of two fixed resistors in a 10:1 divider arrangement, but using a potentiometer makes it easy to set the amplifier's maximum gain. When powered from a vehicle's electrical system, the amplifier's +12 V power source requires filter L1 to eliminate alternator whine. The LM383 can be mounted directly on the heatsink because the mounting tab is at ground potential.

TV AUDIO AMPLIFIER

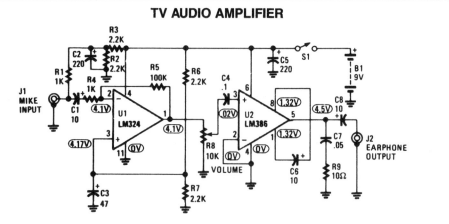

HANDS-ON ELECTRONICS/POPULAR ELECTRONICS

Fig. 1-9

The amplifier picks up the TV's audio output signal and amplifies it to drive a set of earphones for private listening. It is built around an LM324 quad op amp and an LM386 low-power audio amplifier. The circuit uses an inexpensive electret microphone element as the pick-up and a set of earphones as the output device.

LOW-POWER AUDIO AMPLIFIER

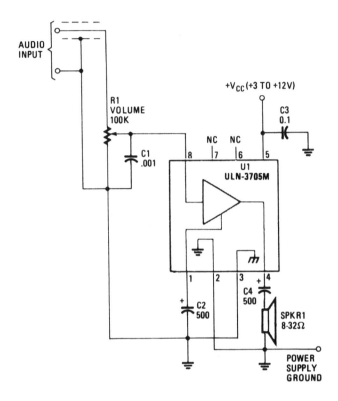

Fig. 1-10

The amplifier operates from supplies ranging up to 12 volts, and operates (with reduced volume) from supply voltages as low as 1.8 volts without having distortion rise to unacceptable levels (its power requirements make it suitable for solar-cell application). Components external to the integrated circuit, U1, consist of four capacitors and a potentiometer for volume control. Capacitor C3 is for decoupling, low-frequency rolloff, and power-supply ripple rejection. Capacitor C4 is an electrolytic type that couples the audio output to an efficient 8- to 32-Ω speaker.

STEREO AMPLIFIER WITH Av. = 200

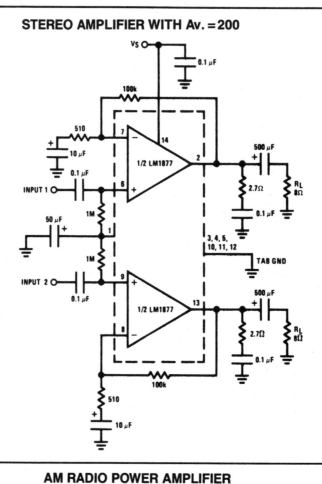

Fig. 1-11

AM RADIO POWER AMPLIFIER

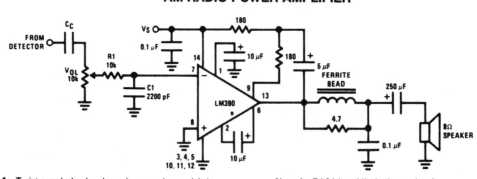

Note 1: Twist supply lead and supply ground very tightly.
Note 2: Twist speaker lead and ground very tightly.
Note 3: Ferrite bead is Ferroxcube K5-001-001/3B with 3 turns of wire.

Note 4: R1C1 band limits input signals.
Note 5: All components must be spaced very close to IC.

Fig. 1-12

AUDIO POWER AMPLIFIER

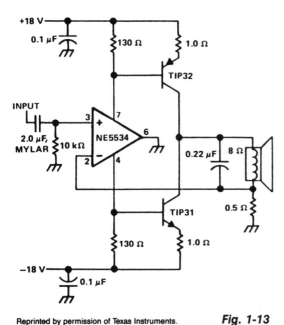

The single speaker amplifier circuit uses current feedback, rather than the more popular voltage feedback. The feedback loop is from the junction of the speaker terminal and a 0.5-Ω resistor, to the inverting input of the NE5534. When the input to the amplifier is positive, the power supply supplies current through the TIP32 and the load to ground. Conversely, with a negative input, the TIP31 supplies current through the load to ground. The gain is set to about 15 (gain = SPKR 8 Ω/0.5 Ω feedback). The 0.22-μF capacitor across the speaker rolls off its response beyond the frequencies of interest. Using the 0.22-μF capacitor specified, the amplifier current output is 3 dB down at 90 kHz where the speaker impedance is about 20 Ω. To set the recommended class A output collector current, adjust the value of either 130-Ω resistor. An output current of 50 to 100 mA will provide a good operating midpoint between the best crossover distortion and power dissipation.

Reprinted by permission of Texas Instruments.   **Fig. 1-13**

6-W AUDIO AMPLIFIER WITH PREAMP

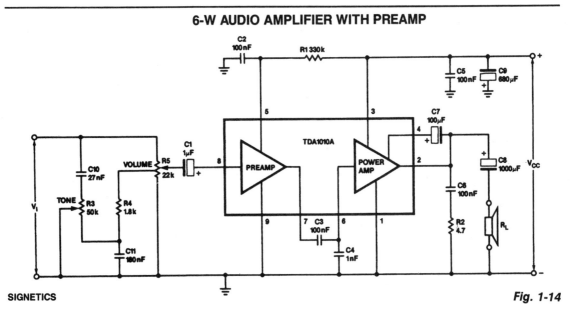

SIGNETICS   *Fig. 1-14*

This monolithic IC, class-B, audio amplifier circuit is a 6-W car radio amplifier for use with 4-Ω and 2-Ω load impedances.

SPLIT-SUPPLY NONINVERTING AMPLIFIER

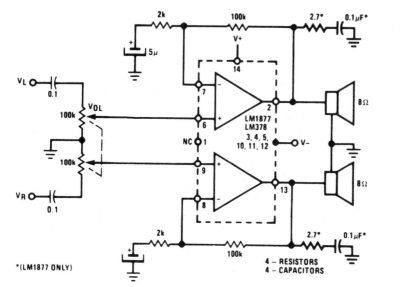

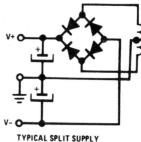

TYPICAL SPLIT SUPPLY

*(LM1877 ONLY)

4 – RESISTORS
4 – CAPACITORS

NATIONAL SEMICONDUCTOR

Fig. 1-15

6-W 8-Ω OUTPUT TRANSFORMERLESS AMPLIFIER

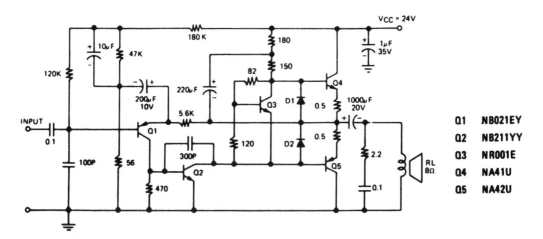

Q1	NB021EY
Q2	NB211YY
Q3	NR001E
Q4	NA41U
Q5	NA42U

NATIONAL SEMICONDUCTOR

Fig. 1-16

2- TO 6-W AUDIO AMPLIFIER WITH PREAMPLIFIER

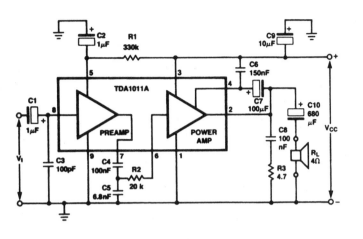

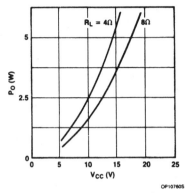

NOTES:
d_{TOT} = 10%; typical values. The available output power is 5% higher when measured at Pin 2 (due to series resistance of C1). *

**Output Power Across R_L
as a Function of Supply
Voltage with Bootstrap**

SIGNETICS

Fig. 1-17

The monolithic integrated audio amplifier circuit is especially designed for portable radio and recorder applications and it delivers up to 4 W in a 4-Ω load impedance.

AUDIO POWER AMPLIFIER

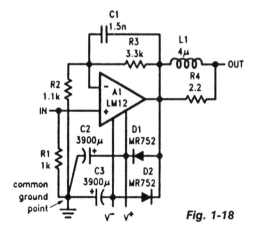

Fig. 1-18

NATIONAL SEMICONDUCTOR CORP.

Output-clamp diodes are mandatory because loudspeakers are inductive loads. Output LR isolation is also used because audio amplifiers are usually expected to handle up to a 2-μF load capacitance. Large, supply-bypass capacitors are located close to the IC so that the rectified load current in the supply leads does not get back into the amplifier, and increase high-frequency distortion. Single-point grounding for all internal leads, plus the signal source and load, is recommended to avoid ground loops, which can increase distortion.

12-W LOW-DISTORTION POWER AMPLIFIER

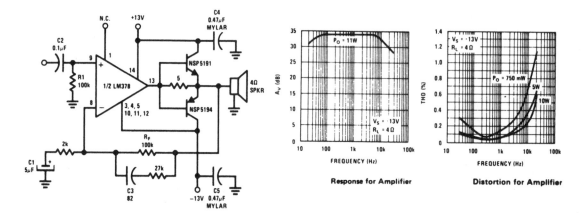

Response for Amplifier

Distortion for Amplifier

Fig. 1-19

10-W POWER AMPLIFIER

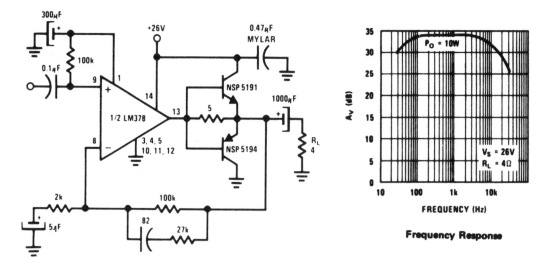

Frequency Response

Fig. 1-20

HIGH SLEW-RATE POWER OP AMP/AUDIO AMP

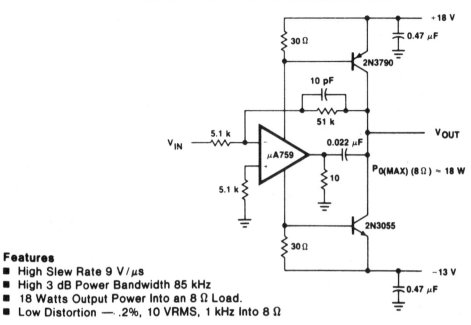

Features
- High Slew Rate 9 V/μs
- High 3 dB Power Bandwidth 85 kHz
- 18 Watts Output Power Into an 8 Ω Load.
- Low Distortion — .2%, 10 VRMS, 1 kHz Into 8 Ω

FAIRCHILD CAMERA

Fig. 1-21

16-W BRIDGE AMPLIFIER

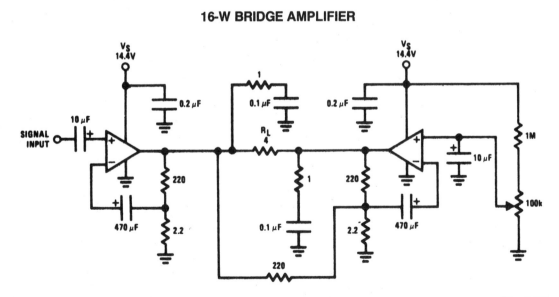

NATIONAL SEMICONDUCTOR

Fig. 1-22

LOW-COST 20-W AUDIO AMPLIFIER

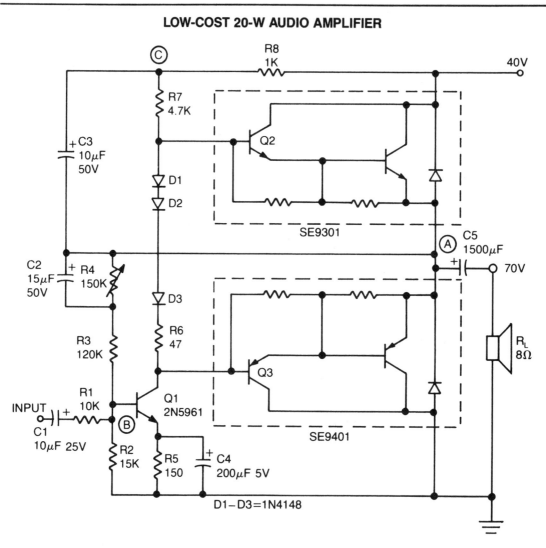

FAIRCHILD CAMERA

Fig. 1-23

This simple inexpensive audio amplifier can be constructed using a couple of TO-220 monolithic Darlington transistors for the push-pull output stage. Frequency response is flat within 1 dB from 30 Hz to 200 kHz with typical harmonic distortion below 0.2%. The amplifier requires only 1.2 V_{rms} for a full 20-W output into an 8-Ω load. Only one other transistor is needed, the TO-92 low-noise high-gain 2N5961 (Q1), to provide voltage gain for driving the output Darlingtons. Its base (point B) is the tie point for ac and dc feedback as well as for the signal input. Input resistance is 10 kΩ. The center voltage at point A is set by adjusting resistor R4. A bootstrap circuit boosts the collector supply voltage of Q1 (point C) to ensure sufficient drive voltage for Q2. This also provides constant voltage across R7, which therefore acts as a current source and, together with diodes D1 to D3, reduces low-signal crossover distortion.

25-WATT AMPLIFIER

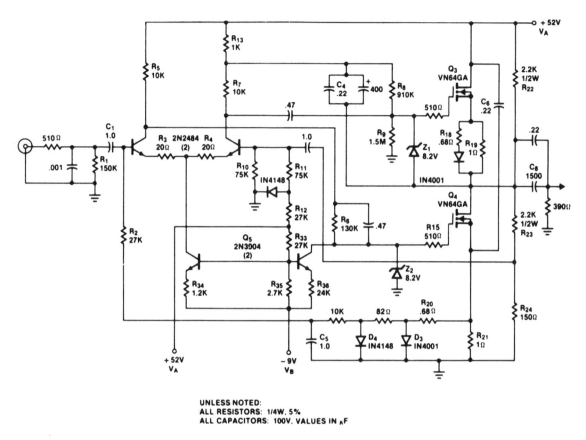

UNLESS NOTED:
ALL RESISTORS: 1/4W, 5%
ALL CAPACITORS: 100V, VALUES IN ₚF

SILICONIX, INC.

Fig. 1-24

Transistors are used for current sources. Base drive for these transistors is derived from main power supply V_A so that their collector current is proportional to the rail voltage. This feature holds the voltage on the diff-amp collectors close to $V_A/2$. The sensitivity of I_Q to V_A is about 3.4 mA/V when V_B is held constant; the sensitivity of I_Q to V_B is -15 mA/V when V_A is held constant. In a practical amplifier with a nonregulated supply, variations in power output will cause fluctuations in V_A, but will not affect V_B; therefore, having I_Q increase slightly with power output will compensate for the 3.4 mA/V $I_Q V_A$ sensitivity. In the case of line voltage variations, since V_A is about five times V_B, the sensitivities cancel and leave a net sensitivity of about 2 mA/V.

50-W AUDIO POWER AMPLIFIER

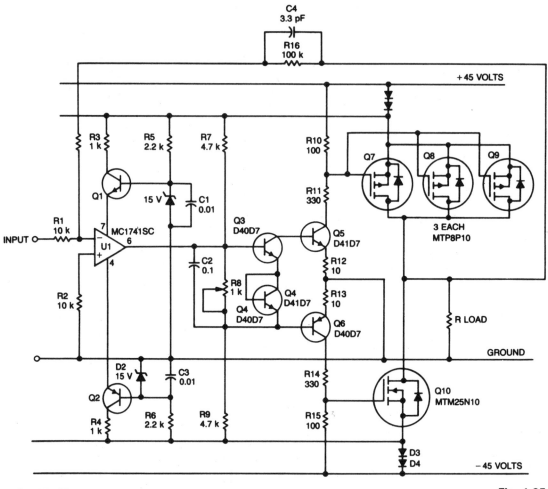

Fig. 1-25

This audio amplifier design approach employs TMOS Power FETs operating in a complementary common-source configuration. They are biased to cutoff, then turn on very quickly when a signal is applied. The advantage of this approach is that the output stage is very stable from a thermal point of view.

U1 is a high slew-rate amp that drives Q3, Q4, and Q6 (operating class AB) providing level transition for the output stage consisting of Q7, Q8, Q9, and Q10. The positive temperature coefficient of the TMOS device enables parallel operation of Q7, Q8, and Q9 and provides a higher power *complementary* device for Q10. These TMOS Power FETs must be driven from a low-source impedance of 100 Ω, in order to actually obtain high turn-on speeds.

75-W AUDIO AMPLIFIER WITH LOAD LINE PROTECTION

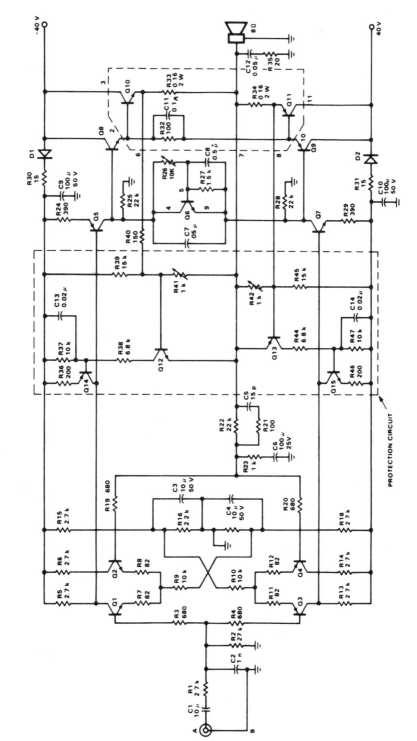

Fig. 1-26

FAIRCHILD CAMERA

90-W AUDIO POWER AMPLIFIER WITH SAFE-AREA PROTECTION

†Put on common heat sink

*34 turns of no. 20 wire on a 3/8" form

**Adjust R6 to set I_Q = 100 mA

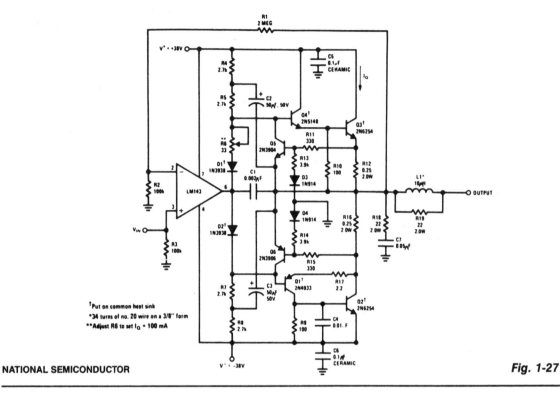

NATIONAL SEMICONDUCTOR

Fig. 1-27

POWER AMPLIFIER

NOTE:
All resistor values are in ohms.

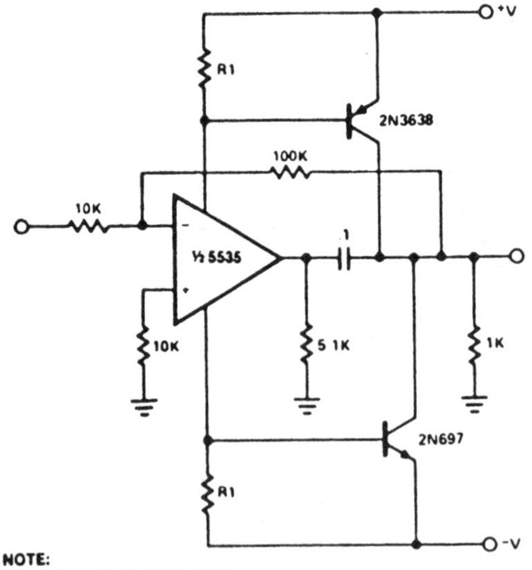

For most applications, the available power from op amps is sufficient. At times, more power-handling capability is necessary. A simple power booster capable of driving moderate loads uses an NE5535 device. Other amplifiers can be substituted only if R1 values are changed because of the I_{CC} current required by the amplifier. R1 should be calculated from the expression:

$$R_1 = \frac{600\ \text{mV}}{I_{CC}}$$

SIGNETICS

Fig. 1-28

CLASS-D POWER AMPLIFIER

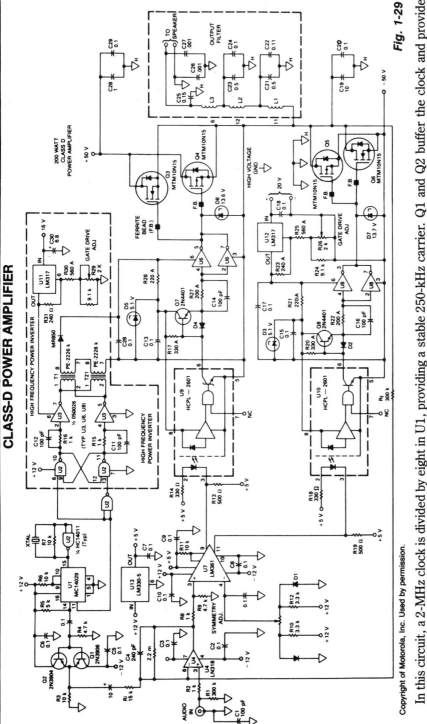

Fig. 1-29

In this circuit, a 2-MHz clock is divided by eight in U1, providing a stable 250-kHz carrier. Q1 and Q2 buffer the clock and provide a low-impedance drive for op amp U4, which is a high-gain amplifier and integrator. U4 accepts audio inputs and converts the 250-kHz square wave into a triangular wave. The summed audio and triangular-wave signal is applied to the input of comparator U7, where it is compared with a dc reference to produce a pulse-width modulated signal at the output of U7.

The output devices switch between the +50 V and −50 V rails in a complementary fashion, driving the output filter that is a sixth-order Butterworth low-pass type, which demodulates the audio and attenuates the carrier and high frequency components. Feedback is provided R_f; amplifier gain is $R_f R_i$.

Specifications: 200 W continuous power into a 4-Ω load; 20 to 20 kHz frequency response +0.5, −1.0 dB at 200 W; THD, IMD 0.5% at 200 W; 1.5-V rms input for rated output; 69 dB S/N ratio, A weighing; 6.6-V ms slew rate.

21

BULL HORN

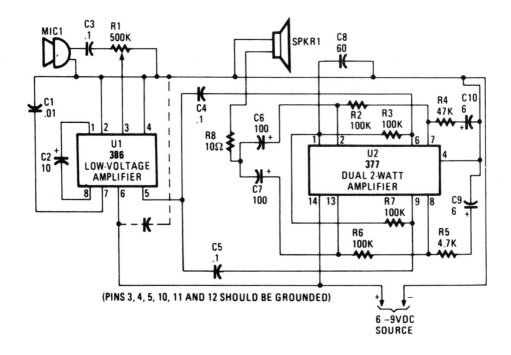

(PINS 3, 4, 5, 10, 11 AND 12 SHOULD BE GROUNDED)

6 –9VDC
SOURCE

Fig. 1-30

The input audio signal is fed to pin 3 of U1, an LM386 low-voltage amplifier, via C3 and R1. Potentiometer R1 sets the drive or volume level. U1, which serves as a driver stage, can be set for a gain of from 20 to 200. The output of U1 at pin 5 is fed to U2—a 377 dual two-watt amplifier connected in parallel to produce about four watts of output power—at pins 6 and 9 via C4 and C5. Frequency stability is determined by R2, R4, and C10 on one side, and the corresponding components R6, R5, and C9 on the other side. The outputs of the two amplifiers (at pins 2 and 13) are capacitively coupled to SPKR1 through C6 and C7.

AUDIO-CIRCUIT BRIDGE LOAD DRIVE

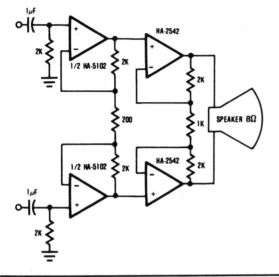

HARRIS

Fig. 1-31

This circuit shows a method which increases the power capability of a drive system for audio speakers. Two HA-2542s are used to operate on half cycles only, which greatly increases their power handling capability. Bridging the speaker, as shown, makes 200 mA of output current available to drive the load. The HA-5102 is used as an ac-coupled, low noise preamplifier, which drives the bridge circuit.

20-W AUDIO AMPLIFIER

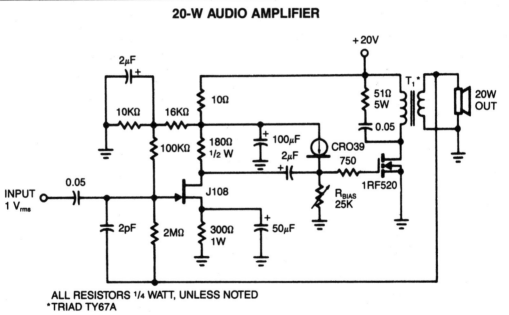

ALL RESISTORS ¼ WATT, UNLESS NOTED
*TRIAD TY67A

SILICONIX

Fig. 1-32

This amplifier delivers 20 W into an 8-Ω load using a single IRF520 driving a transformer coupled output stage. This circuit is similar to the audio output stage used in many inexpensive radios and phonographs. Distortion is less than 5% at 10 W, using very little feedback (3%), with the IRF520 biased at 3 A.

MICRO-SIZED AMPLIFIER

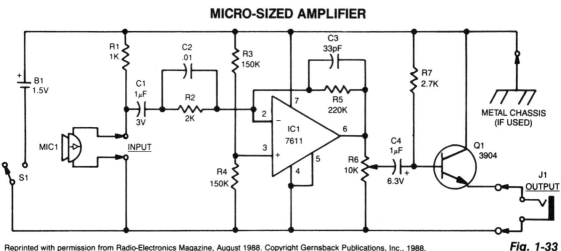

Reprinted with permission from Radio-Electronics Magazine, August 1988. Copyright Gernsback Publications, Inc., 1988.

Fig. 1-33

Sound detected by electret microphone MIC1 is fed to IC1's input through resistor R2, and capacitors C1 and C2. Resistors R2 and R5 determine the overall stage gain, while C2 partially determines the amplifier's frequency response. To ensure proper operation, use a single-ended power supply. R3 and R4 simulate a null condition equal to half the power supply's voltage at IC1's noninverting input. The output of IC1 is transferred to emitter-follower amplifier Q1 via volume control R6. The high-Z-in/low-Z-out characteristic of the emitter-follower matches the moderately high-impedance output of IC1 to a low-impedance headphone load.

AUDIO AMPLIFIER

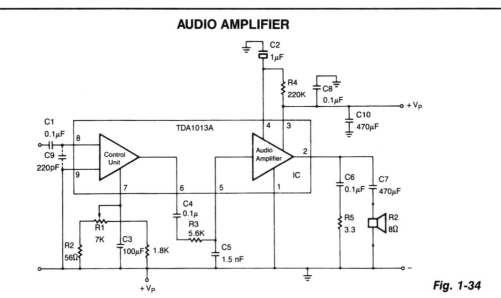

SIGNETICS

Fig. 1-34

C9 is necessary to filter-out rf input interferences. R3 in combination with C5 is used to limit the af frequency bandwidth. The 470-μF power supply decoupling capacitor is C10.

BRIDGE AUDIO POWER AMPLIFIER

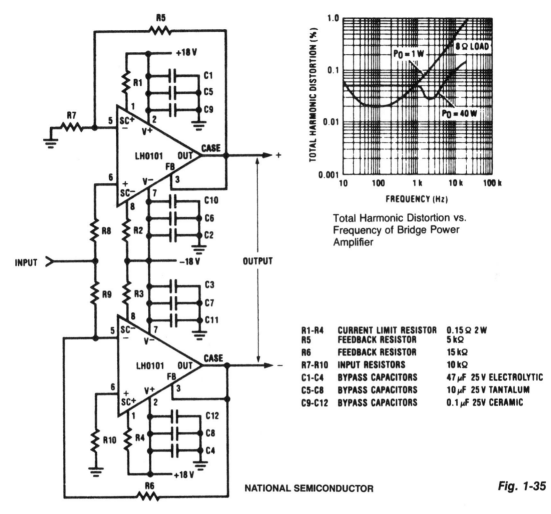

Total Harmonic Distortion vs. Frequency of Bridge Power Amplifier

R1-R4	CURRENT LIMIT RESISTOR	0.15 Ω 2 W
R5	FEEDBACK RESISTOR	5 kΩ
R6	FEEDBACK RESISTOR	15 kΩ
R7-R10	INPUT RESISTORS	10 kΩ
C1-C4	BYPASS CAPACITORS	47 µF 25 V ELECTROLYTIC
C5-C8	BYPASS CAPACITORS	10 µF 25 V TANTALUM
C9-C12	BYPASS CAPACITORS	0.1 µF 25V CERAMIC

NATIONAL SEMICONDUCTOR

Fig. 1-35

PHONO AMPLIFIER

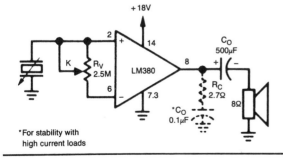

*For stability with high current loads

This circuit is used when maximum input impedance is required or the signal attenuation of the voltage-divider volume control is undesirable.

NATIONAL SEMICONDUCTOR

Fig. 1-36

LINE-OPERATED AMPLIFIER

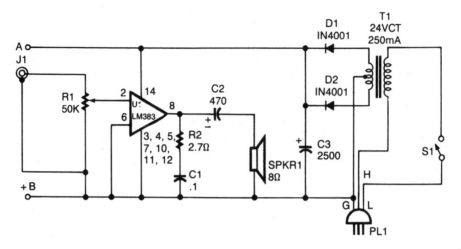

POPULAR ELECTRONICS

Fig. 1-37

T1 isolates the unit from the line, and has a 24-V center-tapped secondary. The output of the transformer is rectified by diodes D1 and D2 and filtered by capacitor C3 to provide 15 to 18 Vdc. The LM383 has built-in protection against speaker shorts.

POWER BOOSTER

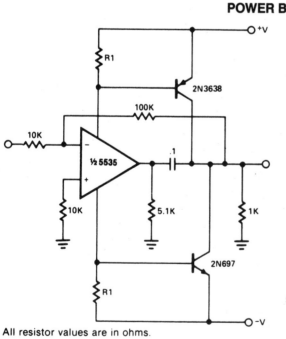

All resistor values are in ohms.

Power booster is capable of driving moderate loads. The circuit as shown uses a NE5535 device. Other amplifiers may be substituted only if R1 values are changed because of the I_{CC} current required by the amplifier. R1 should be calculated from the following expression:

$$R_1 = \frac{600 \text{ mW}}{I_{CC}}$$

SIGNETICS

Fig. 1-38

MINI-STEREO

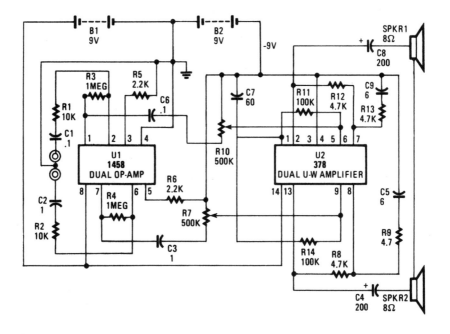

Fig. 1-39

This circuit is built around two chips: the MC1458 dual op amp, configured as a preamplifier, and the LM378 dual 4-watt amplifier. The gain of the preamp is given by R3/R1 for one side and R4/R2 for the other side, which is about 100. That gain can be varied by increasing the ratios. The left and right channel inputs are applied to pins 2 and 6. The left and right outputs of U1 at pins 7 and 2 are coupled through C5/R10 and C3/R6, respectively, to U2 to drive the two 8-Ω loudspeakers.

2

Audio Signal Amplifiers

The sources of the following circuits are contained in the Sources section, which begins on page 182. The figure number contained in the box of each circuit correlates to the source entry in the Sources section.

BALANCE AND LOUDNESS AMPLIFIER

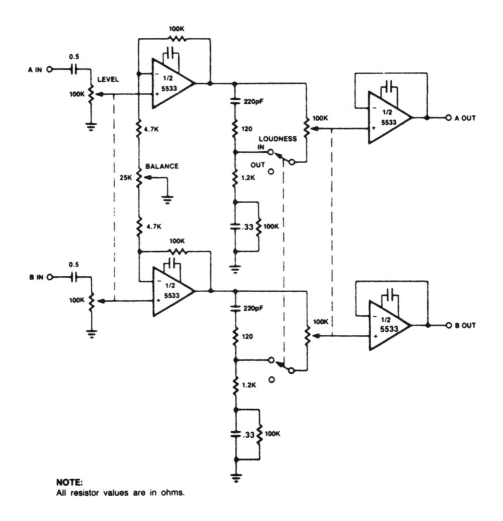

NOTE:
All resistor values are in ohms.

SIGNETICS

Fig. 2-1

This circuit shows a combination of balance and loudness controls. As a result of the nonlinearity of the human hearing system, the low frequencies must be boosted at low listening levels. Balance, level, and loudness controls provide all the listening controls to produce the desired music response.

29

STEREO PREAMPLIFIER

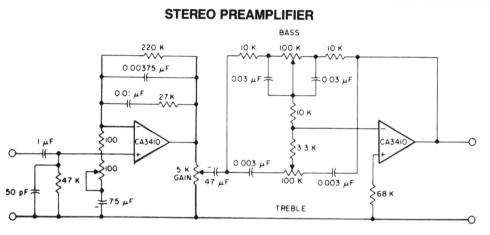

GENERAL ELECTRIC/RCA

Fig. 2-2

This circuit has RIAA playback equalization, tone controls, and adequate gain to drive a majority of commercial power amplifiers, using the CA3410 BiMOS op amp. Total harmonic distortion, when driven to provide a 6-V output, is less than 0.035% in the audio-frequency range of 150 Hz to 40 kHz. A complete stereo preamplifier consists of duplicating this circuit with the two remaining CA3410 amplifiers.

MICROPHONE PREAMPLIFIER

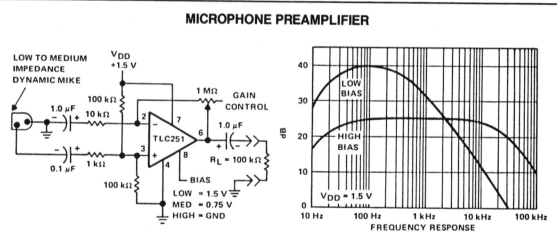

TEXAS INSTRUMENTS

Fig. 2-3

A microphone preamplifier using a :om CMOS op amp, complete with its own battery, is small enough to be put in a small mike case. The amplifier operates from a 1.5-V mercury cell at low supply currents. This preamplifier will operate at very low power levels and maintain a reasonable frequency response as well. The TLC251 operated in the low-bias mode (operating at 1.5 V) draws a supply current of only 10 μA and has a −3-dB frequency response of 27 Hz to 4.8 kHz. With pin 8 grounded, which is designated as the high-bias condition, the upper limit increases to 25 kHz. Supply current is 30 μA under those conditions.

TRANSISTOR HEADPHONE AMPLIFIER

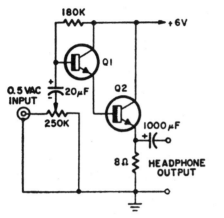

RADIO-ELECTRONICS

Fig. 2-4

STEREO PREAMPLIFIER

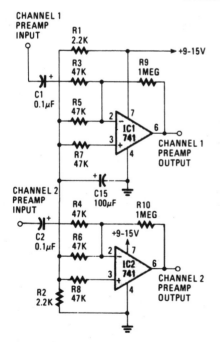

The circuit provides better than 20-dB gain in each channel. A better op-amp type will give a better noise figure and bandpass. In this circuit, the roll-off is acute at 20,000 Hz.

HANDS-ON ELECTRONICS

Fig. 2-5

TRANSFORMERLESS (BALANCED INPUTS) MICROPHONE PREAMP

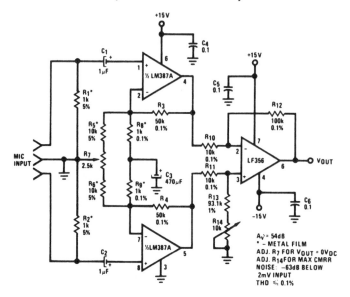

$A_V = 54dB$
* – METAL FILM
ADJ. R_7 FOR $V_{OUT} = 0V_{DC}$
ADJ. R_{14} FOR MAX CMRR
NOISE: –63dB BELOW
2mV INPUT
THD ≤ 0.1%

Fig. 2-6

TRANSFORMERLESS MICROPHONE PREAMPS
(UNBALANCED INPUTS)

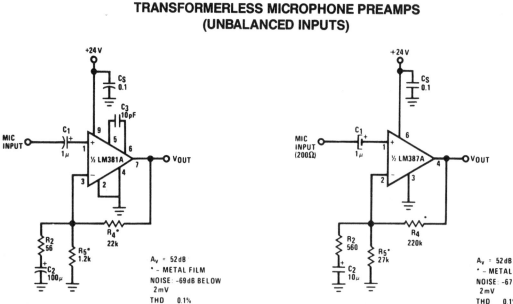

$A_V = 52dB$
* – METAL FILM
NOISE: –69dB BELOW
2mV
THD 0.1%

(a) LM381A S. E. Bias

$A_V = 52dB$
* – METAL FILM
NOISE: –67dB BELOW
2mV
THD 0.1%

(b) LM387A

Fig. 2-7

AUDIO COMPRESSOR

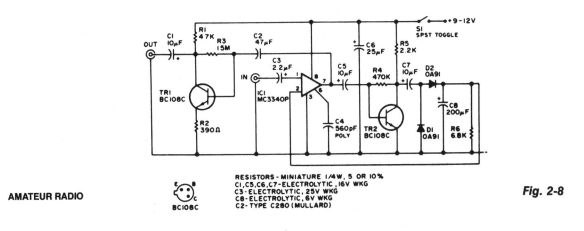

RESISTORS - MINIATURE 1/4W, 5 OR 10%
C1,C5,C6,C7-ELECTROLYTIC, 16V WKG
C3-ELECTROLYTIC, 25V WKG
C8-ELECTROLYTIC, 6V WKG
C2-TYPE C280 (MULLARD)

Fig. 2-8

A MC3340P is used as a variable gain amplifier. TR2's output is rectified control the gain of IC1.

MICROPOWER HIGH-INPUT-IMPEDANCE 20-dB AMPLIFIER

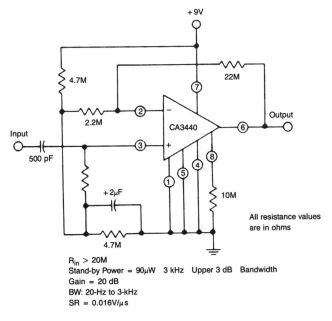

$R_{in} > 20M$
Stand-by Power = 90μW   3 kHz   Upper 3 dB   Bandwidth
Gain = 20 dB
BW: 20-Hz to 3-kHz
SR = 0.016V/μs

Fig. 2-9

This circuit takes advantage of low power drain, high input impedance, and the excellent frequency capability of the CA3440. Only a 500-pF input coupling capacitor is needed to achieve a 20 Hz, −3-dB low-frequency response.

GENERAL-PURPOSE PREAMPLIFIER

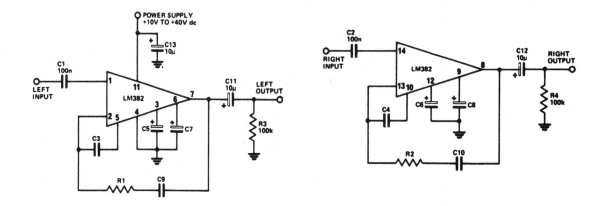

FUNCTION	C3, 4	C5, 6	C7, 8	C9, 10	R1, 2
Phono preamp (RIAA)	330n	10μF	10μF	1n5	1k
Tape preamp (NAB)	68n	10μF	10μF	–	–
Flat 40dB gain	–	–	10μF	–	–
Flat 55dB gain	–	10μF	–	–	–
Flat 80dB gain	–	10μF	10μF	–	–

Fig. 2-10

Not much can be said about how the LM382 works as most of the circuitry is contained within the IC. Most of the frequency-determining components are on the chip—only the capacitors are mounted externally. The LM382 has the convenient characteristic of rejecting ripple on the supply line by about 100 dB, thus greatly reducing the quality requirement for the power supply.

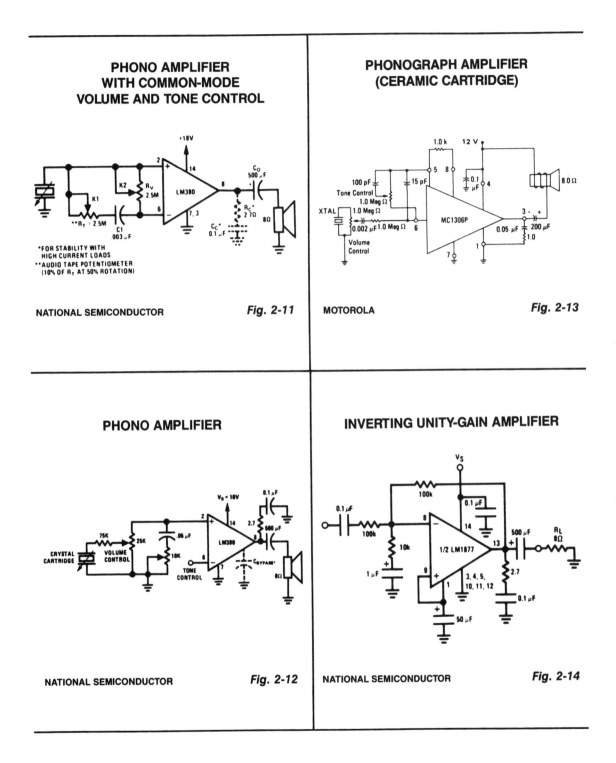

PHONO AMPLIFIER WITH COMMON-MODE VOLUME AND TONE CONTROL

+18V

2
14
+
LM380
8
6
-
7, 3

K2
R_V
2.5M

K1

C_O
500 μF

R_C^*
2.7Ω

8Ω

**R_T = 2.5M
C1
003 μF

C_C^*
0.1 μF

*FOR STABILITY WITH
HIGH CURRENT LOADS
**AUDIO TAPE POTENTIOMETER
(10% OF R_T AT 50% ROTATION)

NATIONAL SEMICONDUCTOR   *Fig. 2-11*

PHONOGRAPH AMPLIFIER (CERAMIC CARTRIDGE)

1.0 k   12 V

5   8

100 pF
Tone Control
1.0 Meg Ω

15 pF

0.1
μF   4

8.0 Ω

XTAL
1.0 Meg Ω
0.002 μF 1.0 Meg Ω   6

MC1306P

3 - +
0.05 μF   200 μF
1.0

7   1

Volume
Control

MOTOROLA   *Fig. 2-13*

PHONO AMPLIFIER

V_S = 18V

2
14
+
LM380
8
7
-

75K
25K
VOLUME
CONTROL

.06 μF

10K

TONE
CONTROL

0.1 μF
2.7
500 μF

C_{BYPASS}^*

8Ω

CRYSTAL
CARTRIDGE

NATIONAL SEMICONDUCTOR   *Fig. 2-12*

INVERTING UNITY-GAIN AMPLIFIER

V_S

100k

0.1 μF

0.1 μF
100k

8
-
14
1/2 LM1877
13
+
9
1

10k

1 μF

3, 4, 5,
10, 11, 12

50 μF

500 μF
+

2.7

0.1 μF

R_L
8Ω

NATIONAL SEMICONDUCTOR   *Fig. 2-14*

35

ULTRA-HIGH-GAIN AUDIO AMPLIFIER

Sometimes called the JFET μ-amp, this circuit provides a very low power, high-gain amplifying function. Since the μ of a JFET increases as drain current decreases, the lower drain current is, the more gain you get. Input dynamic range is sacrificed with increasing gain, however.

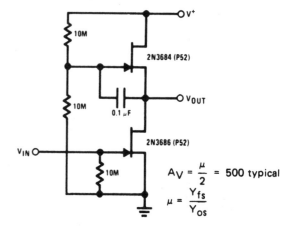

$$A_V = \frac{\mu}{2} = 500 \text{ typical}$$

$$\mu = \frac{Y_{fs}}{Y_{os}}$$

NATIONAL SEMICONDUCTOR

Fig. 2-15

MICROPHONE AMPLIFIER

This circuit operates from a 1.5-Vdc source.

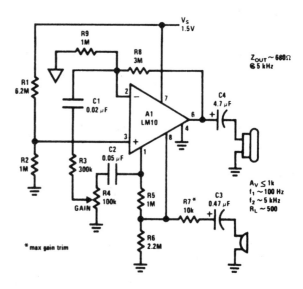

NATIONAL SEMICONDUCTOR

Fig. 2-16

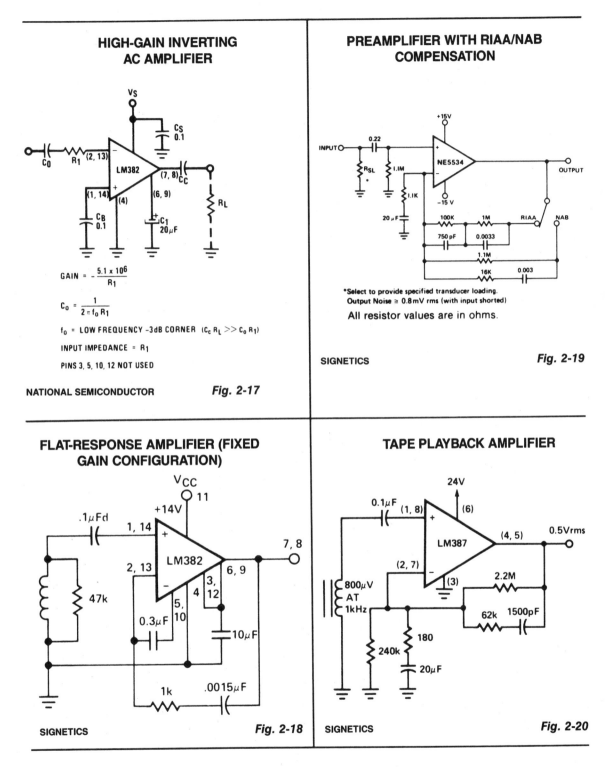

HIGH-GAIN INVERTING AC AMPLIFIER

$$\text{GAIN} = -\frac{5.1 \times 10^6}{R_1}$$

$$C_0 = \frac{1}{2\pi f_0 R_1}$$

f_0 = LOW FREQUENCY –3dB CORNER $(C_c R_L \gg C_0 R_1)$

INPUT IMPEDANCE = R_1

PINS 3, 5, 10, 12 NOT USED

NATIONAL SEMICONDUCTOR   *Fig. 2-17*

PREAMPLIFIER WITH RIAA/NAB COMPENSATION

*Select to provide specified transducer loading.
Output Noise ≥ 0.8mV rms (with input shorted)

All resistor values are in ohms.

SIGNETICS   *Fig. 2-19*

FLAT-RESPONSE AMPLIFIER (FIXED GAIN CONFIGURATION)

SIGNETICS   *Fig. 2-18*

TAPE PLAYBACK AMPLIFIER

SIGNETICS   *Fig. 2-20*

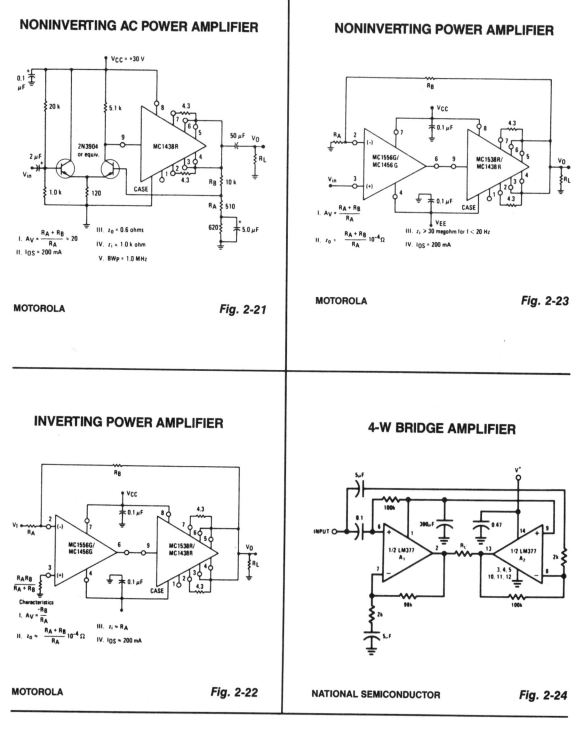

NONINVERTING AC POWER AMPLIFIER

$V_{CC} = +30$ V

20 k
5.1 k

0.1 μF

2N3904
or equiv.

MC1438R

8
7
6
5
4.3

9

3
4
2
1
4.3

CASE

50 μF V_0

R_L

R_B 10 k

R_A 510

2 μF

V_{in}

1.0 k 120

620 5.0 μF

I. $A_V = \dfrac{R_A + R_B}{R_A} \approx 20$

II. $I_{OS} = 200$ mA

III. $z_0 = 0.6$ ohms

IV. $z_i \approx 1.0$ k ohm

V. $BW_P = 1.0$ MHz

MOTOROLA *Fig. 2-21*

NONINVERTING POWER AMPLIFIER

R_B

V_{CC}

R_A 2

$(-)$

MC1556G/
MC1456 G

7

0.1 μF

6 9

MC1538R/
MC1438 R

8
7
6
5
4.3

V_{in} 3

$(+)$

4

0.1 μF

V_{EE}

4
3
2
1
4.3

CASE

V_0

R_L

I. $A_V = \dfrac{R_A + R_B}{R_A}$

II. $z_0 \approx \dfrac{R_A + R_B}{R_A} 10^{-4} \Omega$

III. $z_i > 30$ megohm for $f < 20$ Hz

IV. $I_{OS} = 200$ mA

MOTOROLA *Fig. 2-23*

INVERTING POWER AMPLIFIER

R_B

V_{CC}

0.1 μF

V_I 2

R_A $(-)$

MC1556G/
MC1456G

7

6 9

MC1538R/
MC1438R

8
7
6
5
4.3

$\dfrac{R_A R_B}{R_A + R_B}$ 3

$(+)$

4

0.1 μF

4
3
2
1
4.3

CASE

V_0

R_L

Characteristics

I. $A_V = \dfrac{-R_B}{R_A}$

II. $z_0 = \dfrac{R_A + R_B}{R_A} 10^{-4} \Omega$

III. $z_i \approx R_A$

IV. $I_{OS} \approx 200$ mA

MOTOROLA *Fig. 2-22*

4-W BRIDGE AMPLIFIER

5 μF

V^+

100k

0.1

300μF

0.47

INPUT

6

1/2 LM377
A_1

1

2

R_L

13

1/2 LM377
A_2

14
9

2k

7

3, 4, 5
10, 11, 12

8

2k

98k

100k

5 F

NATIONAL SEMICONDUCTOR *Fig. 2-24*

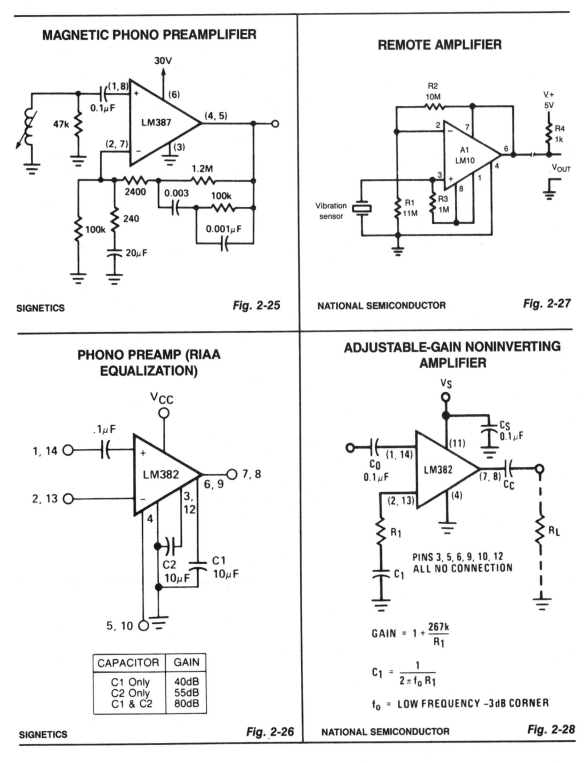

MAGNETIC PHONO PREAMPLIFIER

30V

(1,8)
0.1µF
(6)

47k

LM387

(4, 5)

(2, 7)

(3)

1.2M

2400

0.003

100k

240

100k

0.001µF

20µF

SIGNETICS

Fig. 2-25

REMOTE AMPLIFIER

R2
10M

2

7

A1
LM10

6

V.+
5V

R4
1k

V_{OUT}

3

+

8

1

4

Vibration
sensor

R1
11M

R3
1M

NATIONAL SEMICONDUCTOR

Fig. 2-27

PHONO PREAMP (RIAA EQUALIZATION)

V_{CC}

.1µF

1, 14

+

LM382

6, 9

7, 8

2, 13

−

4

3,
12

C2
10µF

C1
10µF

5, 10

CAPACITOR	GAIN
C1 Only	40dB
C2 Only	55dB
C1 & C2	80dB

SIGNETICS

Fig. 2-26

ADJUSTABLE-GAIN NONINVERTING AMPLIFIER

V_S

C_0
0.1µF

(1, 14)

LM382

(11)

C_S
0.1µF

(7, 8)

C_C

(2, 13)

(4)

R_1

C_1

R_L

PINS 3, 5, 6, 9, 10, 12
ALL NO CONNECTION

$$GAIN = 1 + \frac{267k}{R_1}$$

$$C_1 = \frac{1}{2 \pi f_0 R_1}$$

f_0 = LOW FREQUENCY −3dB CORNER

NATIONAL SEMICONDUCTOR

Fig. 2-28

VOLUME, BALANCE, LOUDNESS, and POWER AMPS

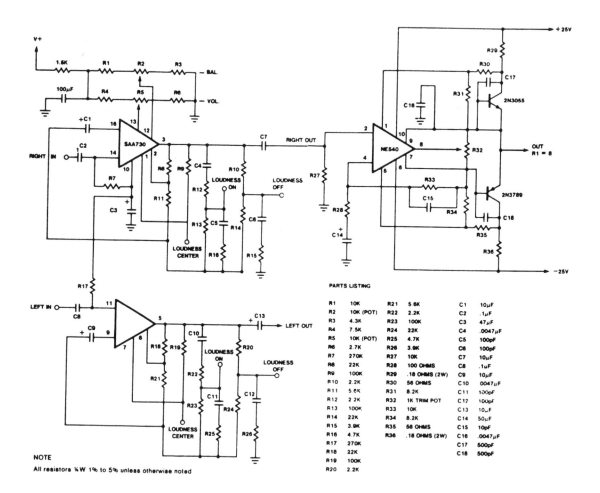

PARTS LISTING

R1	10K	R21	5.6K	C1	10μF
R2	10K (POT)	R22	2.2K	C2	.1μF
R3	4.3K	R23	100K	C3	47μF
R4	7.5K	R24	22K	C4	.0047μF
R5	10K (POT)	R25	4.7K	C5	100pF
R6	2.7K	R26	3.9K	C6	100pF
R7	270K	R27	10K	C7	10μF
R8	22K	R28	100 OHMS	C8	.1μF
R9	100K	R29	.18 OHMS (2W)	C9	10μF
R10	2.2K	R30	56 OHMS	C10	.0047μF
R11	5.6K	R31	8.2K	C11	100pF
R12	2.2K	R32	1K TRIM POT	C12	100pF
R13	100K	R33	10K	C13	10μF
R14	22K	R34	8.2K	C14	50μF
R15	3.9K	R35	56 OHMS	C15	10pF
R16	4.7K	R36	.18 OHMS (2W)	C16	.0047μF
R17	270K			C17	500pF
R18	22K			C18	500pF
R19	100K				
R20	2.2K				

NOTE

All resistors ¼W 1% to 5% unless otherwise noted

SIGNETICS

Fig. 2-29

This circuit should prove suitable as a design example for audio sound application.

AGC WITH SQUELCH CONTROL

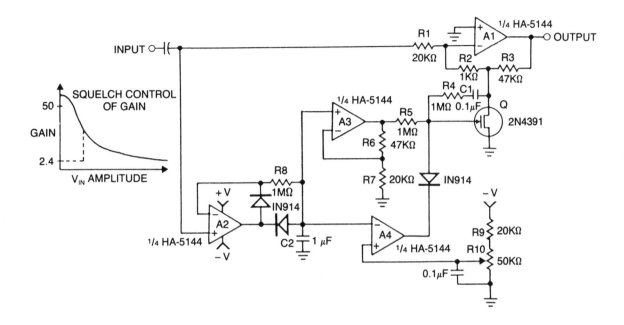

HARRIS

Fig. 2-30

Automatic gain control is a very useful feature in a number of audio amplifier circuits: tape recorders, telephone speaker phones, communication systems and PA systems. This circuit consists of a HA-5144 quad op amp and a FET transistor used as a voltage-controlled resistor to implement an AGC circuit with squelch control. The squelch function helps eliminate noise in communications systems when no signal is present and allows remote hands-free operation of tape recorder systems. Amplifier A1 is placed in an inverting-gain T configuration in order to provide a fairly wide gain range and a small signal level across the FET. The small signal level and the addition of resistors R5 and R6 help reduce nonlinearities and distortion. Amplifier A2 acts as a negative peak detector to keep track of signal amplitude. Amplifier A3 can be used to amplify this peak signal if the cutoff voltage of the FET is higher than desired. Amplifier A4 acts as a comparator in the squelch control section of the circuit. When the signal level falls below the voltage set by R10, the gate of the FET is pulled low—turning it off completely—and reducing the gain to 2.4. The output A4 can also be used as a control signal in applications, such as a hands-free tape recorder system.

AUDIO BOOSTER

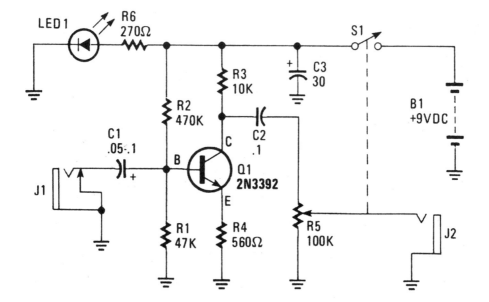

Fig. 2-31

The amplifier's gain is nominally 20 dB. Its frequency response is determined primarily by the value of just a few components—primarily C1 and R1. The values of the schematic diagram provide a response of ± 3.0 dB from about 120 Hz to better than 20 kHz. Actually, the frequency response is ruler flat from about 170 Hz to well over 20 kHz; it's the low end that deviates from a flat frequency response. The low end's roll-off is primarily a function of capacitor C1 (since R1's resistive value is fixed). If C1's value is changed to 0.1 μF, the low end's corner frequency—the frequency at which the low-end roll-off starts—is reduced to about 70 Hz. If you need an even deeper low-end roll-off, change C1 to a 1.0-μF capacitor; if it's an electrolytic type, make certain that it's installed into the circuit with the correct polarity, with the positive terminal connected to Q1's base terminal.

GAIN-CONTROLLED STEREO AMPLIFIER

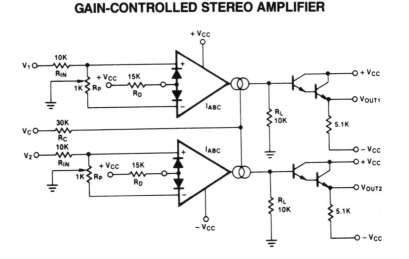

SIGNETICS

Fig. 2-32

MICROPHONE AMPLIFIER

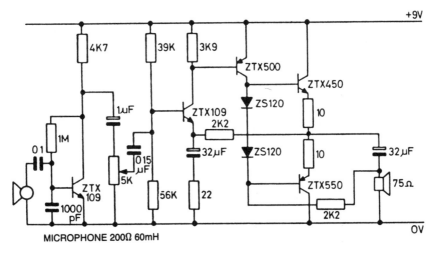

MICROPHONE 200Ω 60mH

ZETEX, formerly FERRANTI

Fig. 2-33

This circuit features the ZTX450/ZTX550 transistors in a push-pull output stage. The following readings were taken at maximum volume:

Input: 0.4 mV rms
Output: 1.8 V rms
Voltage gain: 4500
Max. output before distortion:
  2.25 V rms – supply current = 3.5 mA

Zero output-supply current: 3.5 mA
Wattage: 0.034 W
Frequency response: 250 Hz to 28 kHz

LOW-DISTORTION AUDIO LIMITER

The level at which the audio limiter comes into action can be set with the Limit Level trimmer potentiometer. When that level is exceeded, the output from the Limiter-Detector half of the op-amp (used as a comparator) turns the LED which causes the resistance of the photoresistor to decrease rapidly. That in turn causes the gain of the Limiter half of the op-amp to decrease. When the signal drops below the desired limiting level, the LED turns off, the resistance of the photoresistor increases, and the gain of the Limiter op-amp returns to its normal level set by the combination of resistors R1 and R2. A dual-polarity power supply (±12 volts is desirable) is needed for the op-amp.

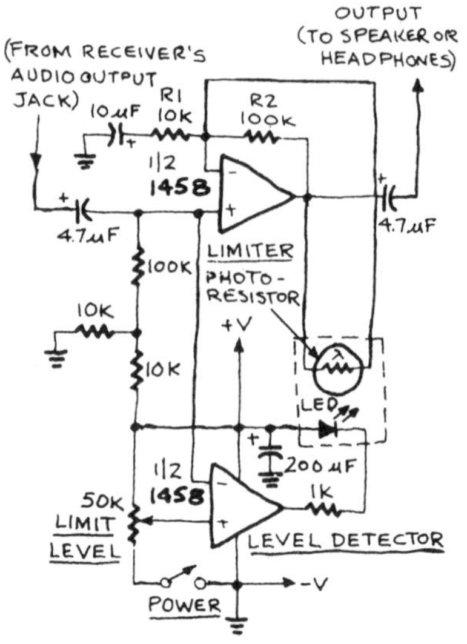

RADIO-ELECTRONICS

Fig. 2-34

SPEECH COMPRESSOR

The amplifier drives the base of a pnp MPS6517 operating common-emitter with a voltage gain of approximately 20. The control R1 varies the quiescent Q point of this transistor so that varying amounts of signal exceed the level V_r. Diode D1 rectifies the positive peaks of Q1's output only when these peaks are greater than $V_r \cong 7.0$ volts. The resulting output is filtered C_x, R_x. R_x controls the charging time constant or attack time. C_x is involved in both charge and discharge. R2 (150 kΩ, input resistance of the emitter-follower Q2) controls the decay time. Making the decay long and attack short is accomplished by making R_x small and R2 large. (A Darlington emitter-follower might be needed if extremely slow decay times are required.) The emitter-follower Q2 drives the AGC Pin 2 of the MC1590 and reduces the gain. R3 controls the slope of signal compression.

MOTOROLA INC.

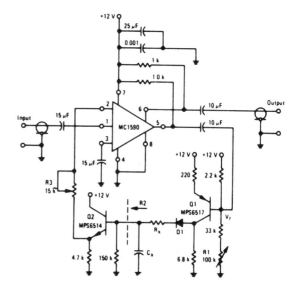

Fig. 2-35

PREAMPLIFIER AND HIGH-TO-LOW IMPEDANCE CONVERTER

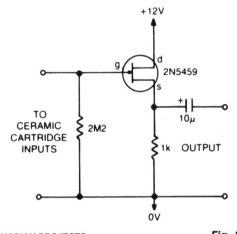

This circuit matches the very high impedance of ceramic cartridges, unity gain, and low-impedance output. By "loading" the cartridge with a 2M2 input resistance, the cartridge characteristics quite closely compensate for the RIAA recording curve. The output from this preamp can be fed to a level pot for mixing.

CANADIAN PROJECTS   *Fig. 2-36*

MAGNETIC PHONO PREAMPLIFIER

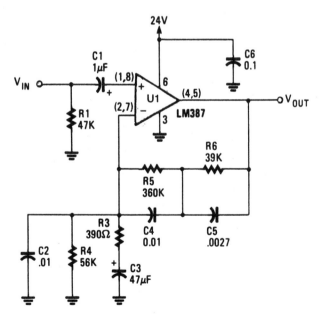

POPULAR ELECTRONICS   *Fig. 2-37*

45

20-dB AUDIO BOOSTER

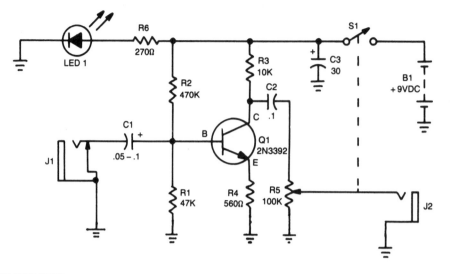

HANDS-ON ELECTRONICS

Fig. 2-38

The amplifier's gain is nominally 20 dB. Its frequency response is determined primarily by the value of just a few components—primarily C1 and R1. The values in the schematic diagram provide a response of ± 3.0 dB from about 120 to over 20,000 Hz. Actually, the frequency response is flat from about 170 to well over 20,000 Hz; it's the low end that deviates from a flat frequency response. The low end's rolloff is primarily a function of capacitor C1, since R1's resistive value is fixed. If C1's value is changed to 0.1 μF, the low end's corner frequency—the frequency at which the low end rolloff starts—is reduced to about 70 Hz. If you need an even deeper low end rolloff, change C1 to a 1.0-μF capacitor. If it's an electrolytic type, make certain that it's installed into the circuit with the correct polarity—with the positive terminal connected to Q1's base terminal.

3

Instrumentation Amplifiers

The sources of the following circuits are contained in the Sources section, which begins on page 182. The figure number contained in the box of each circuit correlates to the source entry in the Sources section.

continued

Bridge Transducer Amplifier
Triple Op-Amp Instrumentation Amplifier
Differential-Input Instrumentation Amplifier with
   High Common-Mode Rejection
Dc-Stabilized Fast Amplifier
Write Amplifier
Linear Amplifiers from CMOS Inverters
Current-Collector Head Amplifier
Instrumentation Amplifier with High Common-
   Mode Rejection
Level-Shifting Isolation Amplifier
Instrumentation Amplifier (Two Op-Amp Design)
Instrumentation Amplifier

Differential-Input Instrumentation Amplifier
High-Impedance Differential Amplifier
High-Speed Instrumentation Amplifier
Very High-Impedance Instrumentation Amplifier
Ac-Coupled Dynamic Amplifier
Forward-Current Booster
RIAA Preamplifier
Professional Audio NAB Tape Playback
   Preamplifier
Automatic Level Control
×1000 Amplifier
Two-Wire to Four-Wire Audio Converter
Low-Power Instrumentation Amplifier

ULTRA-PRECISION INSTRUMENTATION AMPLIFIER

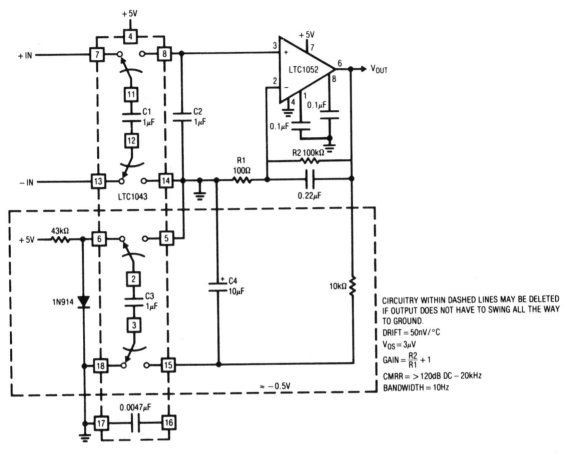

LINEAR TECHNOLOGY

Fig. 3-1

This circuit will run from a single 5 V power supply. The LTC1043 switched-capacitor instrumentation building block provides a differential-to-single-ended transition using a flying-capacitor technique. C1 alternately samples the differential input signal and charges ground referred C2 with this information. The LTC1052 measures the voltage across C2 and provides the circuit's output. Gain is set by the ratio of the amplifier's feedback resistors. Normally, the LTC1052's output stage can swing within 15 mV of ground. If operation all the way to zero is required, the circuit shown in dashed lines can be employed. This configuration uses the remaining LTC1043 section to generate a small negative voltage by inverting the diode drop. This potential drives the 10-KΩ, pull-down resistor, forcing the LTC1052's output into class A operation for voltages near zero. Note that the circuit's switched-capacitor front-end forms a sampled-data filter allowing the common-mode rejection ratio to remain high, even with increasing frequency. The 0.0047 μF unit sets front-end switching frequency at a few hundred Hz.

INSTRUMENTATION AMPLIFIER

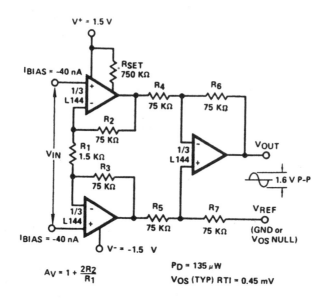

$$A_V = 1 + \frac{2R_2}{R_1}$$

$P_D = 135 \, \mu W$

$V_{OS} \, (TYP) \, RTI = 0.45 \, mV$

SILICONIX

Fig. 3-2

This three-amplifier circuit consumes only 135 μW from a ± 1.5-V power supply. With a gain of 101, the instrumentation amplifier is ideal in sensor interface and biomedical preamplifier applications. The first stage provides all of the gain, and the second stage is used to provide common-mode rejection and double-ended to single-ended conversion.

ISOLATION AMPLIFIER FOR MEDICAL TELEMETRY

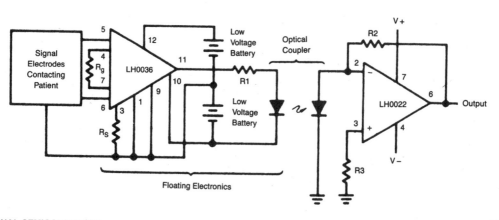

NATIONAL SEMICONDUCTOR

Fig. 3-3

STRAIN GAUGE INSTRUMENTATION AMPLIFIER

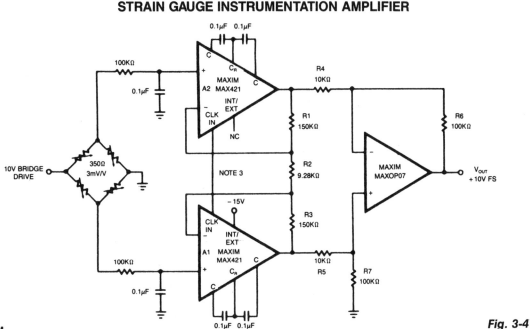

Fig. 3-4

This circuit has an overall gain of 320. More gain can be obtained by lowering the value of R2. Untrimmed V_{OS} is 10 μV, and V_{OS} tempco is less than 0.1 μV/°C. In many circuits, the OP07 can be omitted, with the two MAX421 differential outputs connected directly to the differential inputs of an integrating a/d.

INSTRUMENTATION AMPLIFIER

LINEAR TECHNOLOGY CORP.

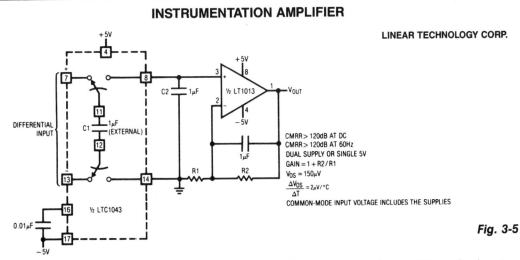

Fig. 3-5

LTC1043 and LT1013 dual op amps are used to create a dual instrumentation amplifier using just two packages. A single DPDT section converts the differential input to a ground-referred single-ended signal at the LT1013's input. The 0.01-μF capacitor at pin 16 sets the switching frequency at 500 Hz.

WIDEBAND INSTRUMENTATION AMPLIFIER

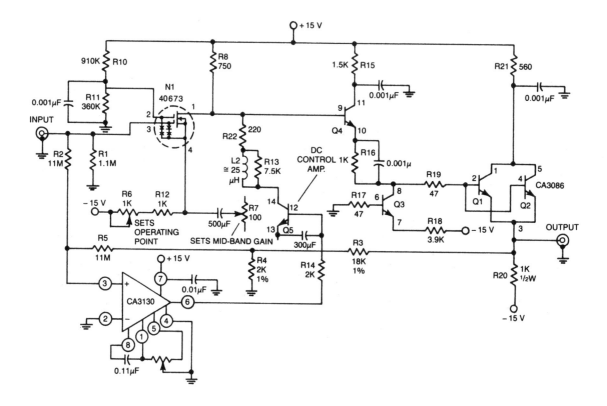

NOTES:
± 15 - VOLT SUPPLIES BYPASSED TO GROUND WITH 5μF CAPACITORS
Q1 - Q5: CA3086 TRANSISTOR—ARRAY IC

GE/RCA

Fig. 3-6

Has an input resistance of 1-MΩ, a bandwidth from dc to about 35 MHz, and a gain of 10 times. Low-frequency gain is provided by a CA3130 BiMOS op amp operated as a single-supply amplifier. High-frequency gain is provided by a 40673 dual-gate MOSFET. The entire amplifier is nulled by shorting the input to ground and adjusting R9 for zero dc output voltage.

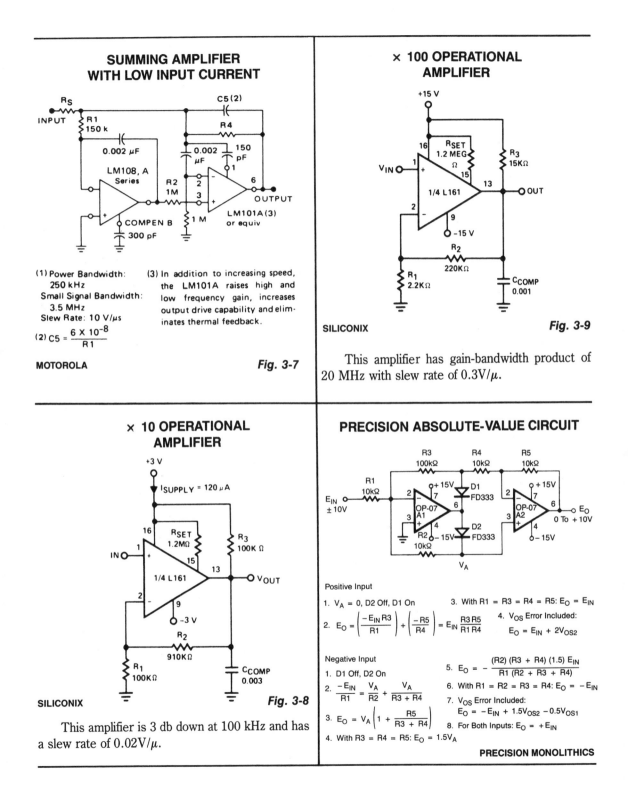

SUMMING AMPLIFIER WITH LOW INPUT CURRENT

(1) Power Bandwidth: 250 kHz
Small Signal Bandwidth: 3.5 MHz
Slew Rate: 10 V/μs

(2) $C5 = \dfrac{6 \times 10^{-8}}{R1}$

(3) In addition to increasing speed, the LM101A raises high and low frequency gain, increases output drive capability and eliminates thermal feedback.

MOTOROLA

Fig. 3-7

× 100 OPERATIONAL AMPLIFIER

SILICONIX

Fig. 3-9

This amplifier has gain-bandwidth product of 20 MHz with slew rate of 0.3V/μ.

× 10 OPERATIONAL AMPLIFIER

SILICONIX

Fig. 3-8

This amplifier is 3 db down at 100 kHz and has a slew rate of 0.02V/μ.

PRECISION ABSOLUTE-VALUE CIRCUIT

Positive Input

1. $V_A = 0$, D2 Off, D1 On

2. $E_O = \left(\dfrac{-E_{IN} R3}{R1}\right) + \left(\dfrac{-R5}{R4}\right) = E_{IN}\dfrac{R3\,R5}{R1\,R4}$

3. With $R1 = R3 = R4 = R5$: $E_O = E_{IN}$

4. V_{OS} Error Included:
   $E_O = E_{IN} + 2V_{OS2}$

Negative Input

1. D1 Off, D2 On

2. $\dfrac{-E_{IN}}{R1} = \dfrac{V_A}{R2} + \dfrac{V_A}{R3 + R4}$

3. $E_O = V_A\left(1 + \dfrac{R5}{R3 + R4}\right)$

4. With $R3 = R4 = R5$: $E_O = 1.5V_A$

5. $E_O = -\dfrac{(R2)(R3 + R4)(1.5)\,E_{IN}}{R1(R2 + R3 + R4)}$

6. With $R1 = R2 = R3 = R4$: $E_O = -E_{IN}$

7. V_{OS} Error Included:
   $E_O = -E_{IN} + 1.5V_{OS2} - 0.5V_{OS1}$

8. For Both Inputs: $E_O = +E_{IN}$

PRECISION MONOLITHICS

53

INSTRUMENTATION AMPLIFIER

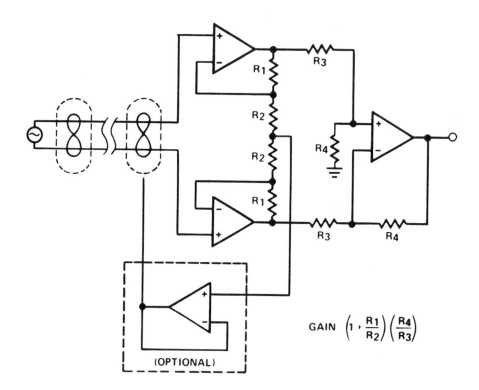

GAIN $\left(1 + \dfrac{R_1}{R_2}\right)\left(\dfrac{R_4}{R_3}\right)$

HARRIS   *Fig. 3-11*

Instrumentation amplifiers (differential amplifiers) are specifically designed to extract and amplify small differential signals from much larger common-mode voltages. To serve as building blocks in instrumentation amplifiers, op amps must have very low offset voltage drift, high gain and wide bandwidth. The HA-4620/5604 is suited for this application. The optional circuitry makes use of the fourth amplifier section as a shield driver which enhances the ac common mode rejection by nullifying the effects of capacitance-to-ground mismatch between input conductors.

CURRENT-COLLECTOR HEAD-AMPLIFIER

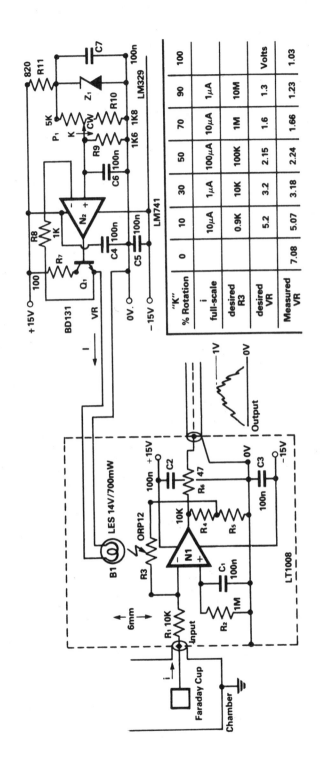

"K" % Rotation	0	10	30	50	70	90	100
i full-scale		10µA	1µA	100µA	10µA	1µA	100
desired R3		0.9K	10K	100K	1M	10M	
desired VR		5.2	3.2	2.15	1.6	1.3	Volts
Measured VR	7.08	5.07	3.18	2.24	1.66	1.23	1.03

Fig. 3-12

ELECTRONIC ENGINEERING

To amplify small-current signals, such as from an electron-collector inside a vacuum chamber, it is convenient for reasons of noise and bandwidth to have a "head-amplifier" attached to the chamber. Op amp N1 is a precision bipolar device with extremely low bias current and offset voltage (1) as well as low noise, which allows the 100:1 feedback attenuator R4:R5. The resistance of R3 can be varied from above 10 MΩ to below 1R, and so the nominal 0 to 1 V-peak output signal corresponds to input current ranges of 1 nA to 10 µA; this current, i, enters via the protective resistor R1. Light from the bulb B1 shines on R3, and the filament current I is controlled by the op amp N2.

The reference voltage VR is "shaped" by the resistors R9R10 so as to tailor the bulb and LDR characteristics to the desired current ranges. Thus, rotation of the calibrated knob K gives the appropriate resistance to R3 for the peak-current scale shown.

INSTRUMENTATION METER DRIVER

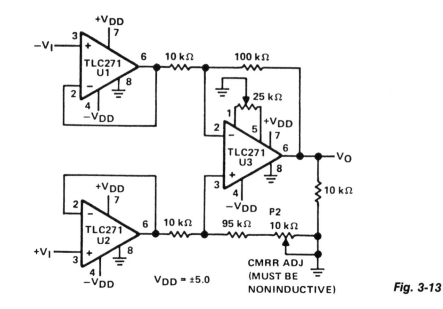

TEXAS INSTRUMENTS

$V_{DD} = \pm 5.0$

Fig. 3-13

Three op amps U1, U2, and U3 are connected in the basic instrumentation amplifier configuration. Operating from ± 5 V, pin 8 of each op amp is connected directly to ground and provides the ac performance desired in this application (high-bias mode). P1 is for offset error correction and P2 allows adjustment of the input common-mode rejection ratio. The high input impedance allows megohms without loading. The resulting circuit frequency response is 200 kHz at −3 dB and has a slew rate of 4.5 V/µs.

SATURATED STANDARD-CELL AMPLIFIER

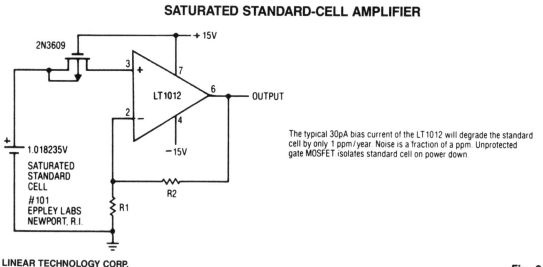

The typical 30pA bias current of the LT1012 will degrade the standard cell by only .1 ppm/year. Noise is a fraction of a ppm. Unprotected gate MOSFET isolates standard cell on power down.

LINEAR TECHNOLOGY CORP.

Fig. 3-14

INSTRUMENT PREAMP

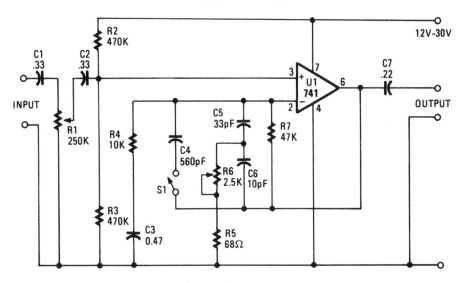

Fig. 3-15

The input impedance is the value of potentiometer R1. If your instrument has extra-deep bass, change capacitor C1 to 0.5 μF. What appears to be an extra part in the feedback loop is a brightening tone control. The basic feedback from the op amp's output (pin 6) to the inverting input (pin 2) consists of resistor R7, and the series connection of resistor R4 and capacitor C3 produces a voltage gain of almost 5 (almost 14 dB). That should be more extra oomph than usually needed. If the circuit is somewhat short on bass response, increase the value of capacitor C3 from 1 to 10 μF. Start with 1 μF and increase the value until you get the bass effect you want.

INSTRUMENTATION AMPLIFIER WITH ± 100-V COMMON-MODE RANGE

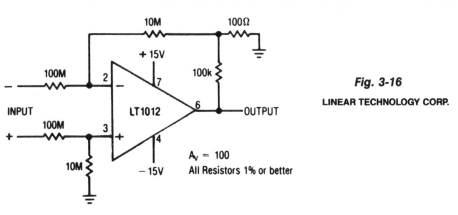

Fig. 3-16

LINEAR TECHNOLOGY CORP.

$A_v = 100$
All Resistors 1% or better

57

VARIABLE-GAIN
DIFFERENTIAL-INPUT INSTRUMENTATION AMPLIFIER

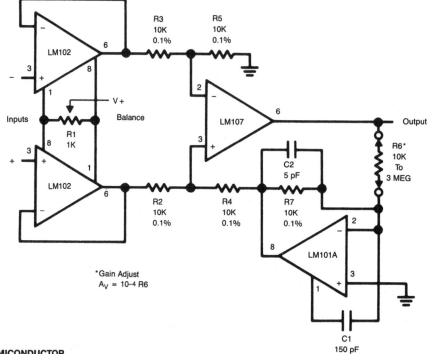

*Gain Adjust
$A_V = 10{-}4 \ R6$

Fig. 3-17

INSTRUMENTATION AMPLIFIER

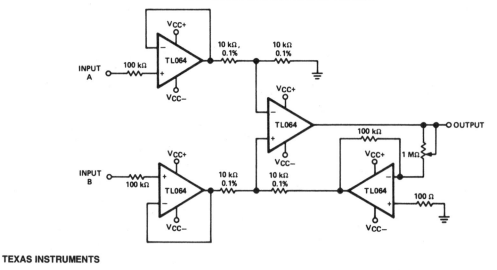

Fig. 3-18

DIFFERENTIAL INSTRUMENTATION AMPLIFIER

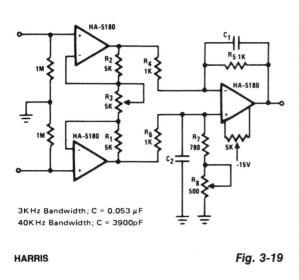

3K Hz Bandwidth; C = 0.053 μF
40K Hz Bandwidth; C = 3900pF

HARRIS

Fig. 3-19

This circuit relies on an extremely high input impedance for effective operation. The HA-5180 with its JFET input stage, performs well as a preamplifier. The standard three amplifier configuration is used with very close matching of the resistor ratios R5/R4 and (R7 + R8)/R6, to ensure high common-mode rejection (CMR). The gain is controlled through R3 and is equal to 2R1/R3. Additional gain can be had by increasing the ratios R5/R4 and (R7 + R8)/R6. The capacitors C1 and C2 improve the ac response by limiting the effects of transients and noise. Two suggested values are given for maximum transient suppression at frequencies of interest. Some of the faster DVM's are operating at peak sampling frequency of 3-kHz, hence the 4-kHz, low-pass time constant. The 40-kHz, low-pass time constant for ac voltage ranges is an arbitrary choice, but should be chosen to match the bandwidth of the other components in the system. C1 and C2 might however, reduce CMR for ac signals if not closely matched. Input impedances have also been added to provide adequate dc bias currents for the HA-5180 when open-circuited.

THERMOCOUPLE PREAMPLIFIER

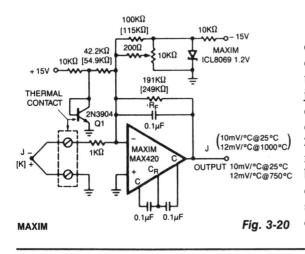

MAXIM

Fig. 3-20

The MAX420 is operated at a gain of 191 to convert the 52 μV/°C output of the type J thermocouple to a 10 mV/°C signal. The −2.2 mV/°C tempco of the 2N3904 is added into the summing junction with a gain of 42.2 to provide cold-junction compensation. The ICL8069 is used to remove the offset caused by the 600-mV initial voltage of the 2N3904. Adjust the 10-KΩ trimpot for the proper reading with the 2N3904 and isothermal connection block at a temperature near the center of the circuit's operating range. Use the component values shown in parentheses when using a type K thermocouple.

BIOMEDICAL INSTRUMENTATION DIFFERENTIAL AMPLIFIER

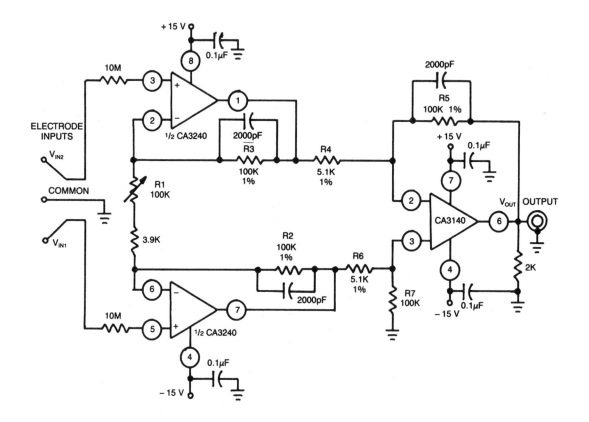

GE/RCA

Fig. 3-21

This differential amplifier uses the isolated high-impedance inputs of the CA3420 BiMOS op amp. Because the CA3240's input current is only 50 pA maximum, 10-MΩ resistors can be used in series with the input probes to limit the current to 2 µA under a fault condition.

LOW-SIGNAL-LEVEL HIGH-
IMPEDANCE INSTRUMENTATION AMPLIFIER

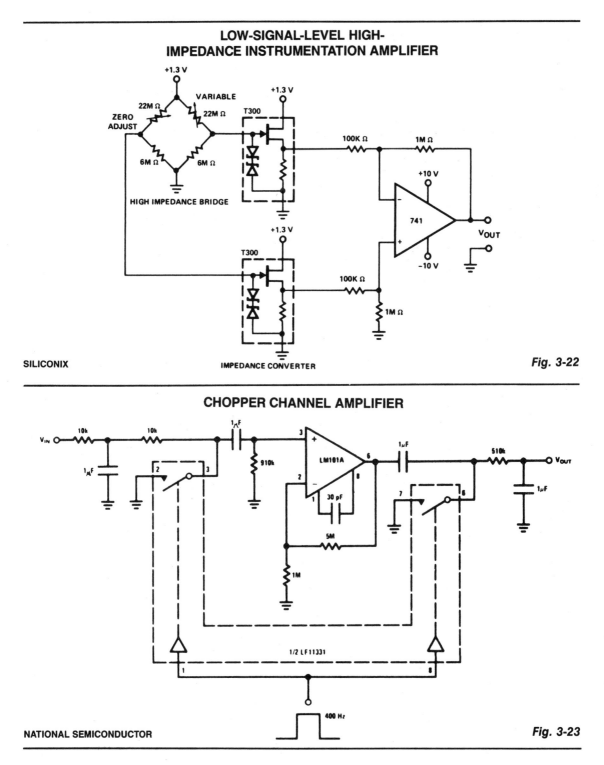

SILICONIX

IMPEDANCE CONVERTER

Fig. 3-22

CHOPPER CHANNEL AMPLIFIER

NATIONAL SEMICONDUCTOR

Fig. 3-23

HIGH-GAIN DIFFERENTIAL INSTRUMENTATION AMPLIFIER

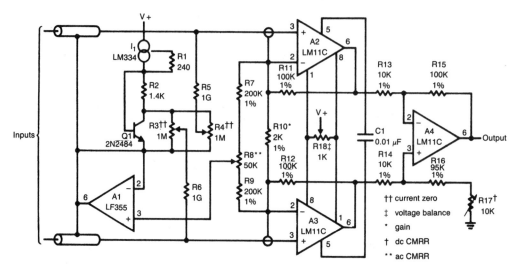

NATIONAL SEMICONDUCTOR

Fig. 3-24

This circuit includes input guarding, cable bootstrapping, and bias-current compensation. Differential bandwidth is reduced by C1, which also makes common-mode rejection less dependent on matching of input amplifiers.

HIGH-IMPEDANCE BRIDGE AMPLIFIER

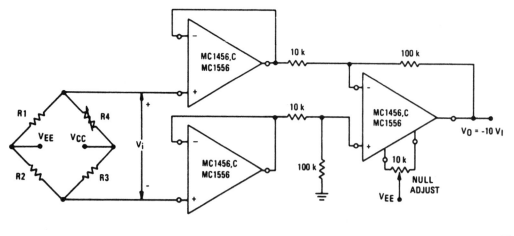

MOTOROLA

Fig. 3-25

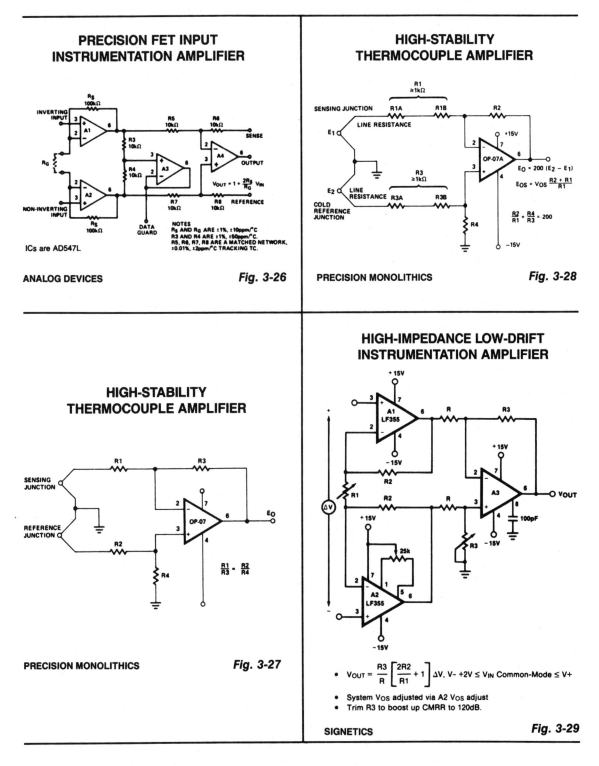

PRECISION FET INPUT INSTRUMENTATION AMPLIFIER

$$V_{OUT} = 1 + \frac{2R_S}{R_G} V_{IN}$$

NOTES
R_S AND R_G ARE ±1%, ±10ppm/°C
R3 AND R4 ARE ±1%, ±50ppm/°C.
R5, R6, R7, R8 ARE A MATCHED NETWORK,
±0.01%, ±2ppm/°C TRACKING TC.

ICs are AD547L

ANALOG DEVICES

Fig. 3-26

HIGH-STABILITY THERMOCOUPLE AMPLIFIER

$E_O = 200 (E_2 - E_1)$

$E_{OS} = V_{OS} \dfrac{R2 + R1}{R1}$

$\dfrac{R2}{R1} = \dfrac{R4}{R3} = 200$

PRECISION MONOLITHICS

Fig. 3-28

HIGH-STABILITY THERMOCOUPLE AMPLIFIER

$$\frac{R1}{R3} = \frac{R2}{R4}$$

PRECISION MONOLITHICS

Fig. 3-27

HIGH-IMPEDANCE LOW-DRIFT INSTRUMENTATION AMPLIFIER

- $V_{OUT} = \dfrac{R3}{R} \left[\dfrac{2R2}{R1} + 1 \right] \Delta V$, V− +2V ≤ V_{IN} Common-Mode ≤ V+

- System V_{OS} adjusted via A2 V_{OS} adjust
- Trim R3 to boost up CMRR to 120dB.

SIGNETICS

Fig. 3-29

63

BATTERY-POWERED BUFFER AMPLIFIER

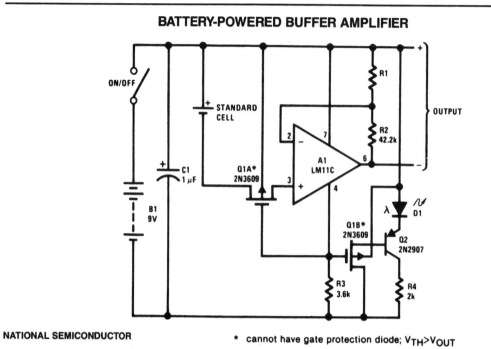

NATIONAL SEMICONDUCTOR

* cannot have gate protection diode; $V_{TH} > V_{OUT}$

Fig. 3-30

This circuit has negligible loading and disconnects the cell for low supply voltage or overload on output. The indicator diode extinguishes as disconnect circuitry is activated.

BRIDGE TRANSDUCER AMPLIFIER

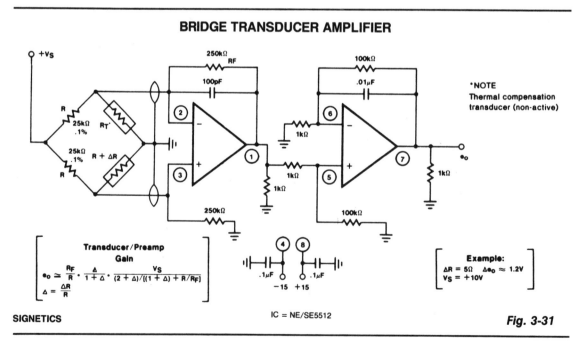

*NOTE
Thermal compensation
transducer (non-active)

$$e_o \simeq \frac{R_F}{R} \cdot \frac{\Delta}{1+\Delta} \cdot \frac{v_S}{(2+\Delta)/[(1+\Delta)+R/R_F]}$$

$$\Delta = \frac{\Delta R}{R}$$

Transducer/Preamp Gain

Example:
$\Delta R = 5\Omega$ $\Delta e_o \approx 1.2V$
$v_S = +10V$

SIGNETICS

IC = NE/SE5512

Fig. 3-31

TRIPLE OP-AMP INSTRUMENTATION AMPLIFIER

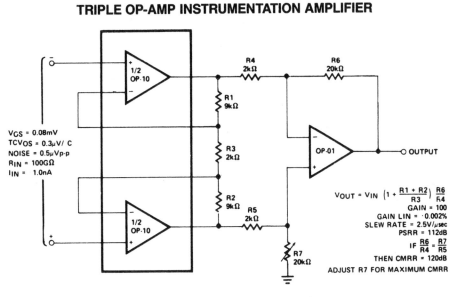

V_{OS} = 0.08mV
TCV_{OS} = 0.3μV/°C
NOISE = 0.5μVp-p
R_{IN} = 100GΩ
I_{IN} = 1.0nA

$$V_{OUT} = V_{IN} \left(1 + \frac{R1 + R2}{R3}\right) \frac{R6}{R4}$$

GAIN = 100
GAIN LIN = ·0.002%
SLEW RATE = 2.5V/μsec
PSRR = 112dB

$$\text{IF } \frac{R6}{R4} = \frac{R7}{R5}$$

THEN CMRR = 120dB
ADJUST R7 FOR MAXIMUM CMRR

PRECISION MONOLITHICS

Fig. 3-32

DIFFERENTIAL-INPUT INSTRUMENTATION AMPLIFIER WITH HIGH COMMON-MODE REJECTION

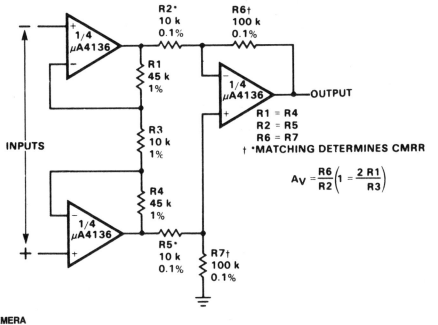

R1 = R4
R2 = R5
R6 = R7
† *MATCHING DETERMINES CMRR

$$A_V = \frac{R6}{R2}\left(1 = \frac{2 \, R1}{R3}\right)$$

FAIRCHILD CAMERA

Fig. 3-33

DC-STABILIZED FAST AMPLIFIER

This amplifier functions over a wide range of gains, typically 1 – 10. It combines the LT1010 and a fast discrete stage with an LT1008 based dc stabilizing loop. Q1 and Q2 form a differential stage which single-ends into the LT1010. The circuit delivers 1 V pk-pk into a typical 75-Ω video load. At $A = 2$, the gain is within 0.5 dB to 10 MHz with the − 3-dB point occurring at 16 MHz. At $A = 10$, the gain is flat (± 0.5 dB to 4 MHz) with a − 3-dB point at 8 MHz. The peaking adjustment should be optimized under loaded output conditions. This is a simple stage for fast applications where relatively low output swing is required. Its 1 V pk-pk output works nicely for video circuits. A possible problem is the relatively high bias current, typically 10 μA. Additional swing is possible, but more circuitry is needed.

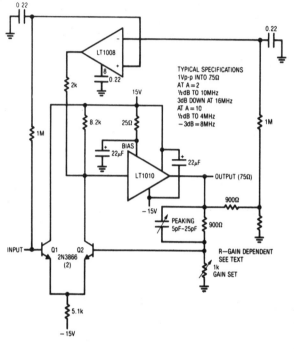

LINEAR TECHNOLOGY CORP.

Fig. 3-34

WRITE AMPLIFIER

The proliferation of industrial and computerized equipment containing programmable memory has increased the need for reliable recording media. The magnetic tape medium is presently one of the most widely used methods. The primary component of any magnetic recording mechanism is the write mechanism. The concept of the write generator is very basic. The digital input causes both a change in the output amplitude, as well as a change in frequency. This type of operation is accomplished by altering the value of a resistor in the standard twin-tee oscillator. A HI-201 analog switch was used to facilitate the switching action. The effect of the external components on the feedback network requires R6A and R6B to be much smaller than would normally have been expected when using the twin-tee feedback scheme.

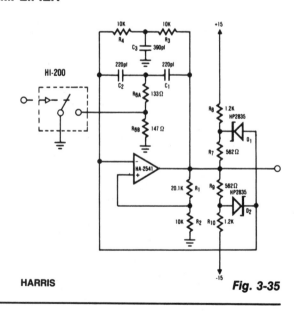

HARRIS

Fig. 3-35

LINEAR AMPLIFIERS FROM CMOS INVERTERS

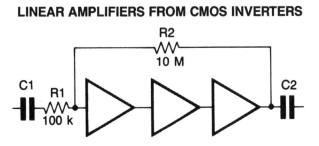

ELECTRONIC ENGINEERING

Fig. 3-36

CMOS inverters can be used as linear amplifiers if negative feedback is applied. Best linearity is obtained with feedback applied around three inverters, which gives almost perfect linearity up to an output swing of 5 V p-p with a 10-V supply rail. The gain is set by the ratio of R1 and R2 and the values shown are typical for a gain of 100. The high-frequency response with the values shown is almost flat to 20 kHz. The frequency response is determined by C1 and C2. This circuit is not suitable for low-level signals because the signal-to-noise ratio is only approx. 50 dB with 5-V p-p output with the values shown.

CURRENT-COLLECTOR HEAD AMPLIFIER

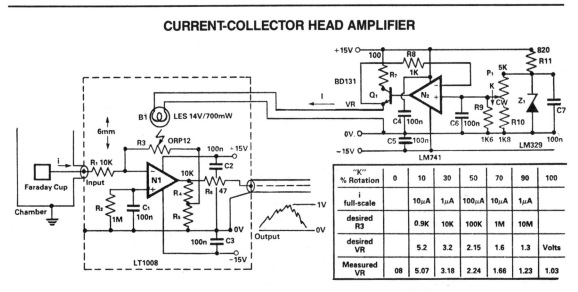

"K" % Rotation	0	10	30	50	70	90	100
i full-scale		10μA	1μA	100μA	10μA	1μA	
desired R3		0.9K	10K	100K	1M	10M	
desired VR		5.2	3.2	2.15	1.6	1.3	Volts
Measured VR	08	5.07	3.18	2.24	1.66	1.23	1.03

ELECTRONIC ENGINEERING

Fig. 3-37

To amplify small-current signals, such as those from an electron-collector inside a vacuum chamber, it is convenient for reasons of noise and bandwidth to have a "head-amplifier" attached to the chamber. Op amp N_1 is a precision bipolar device with extremely low bias current and offset voltage (1) as well as low noise, which allows the 100:1 feedback attenuator to be employed. The resistance of R_3 can be varied from above 10 MΩ to below 1 kΩ, and so the nominal 0 to 1 V-peak output signal corresponds to input current ranges of 1 nA to 10 μA.

INSTRUMENTATION AMPLIFIER
WITH HIGH COMMON-MODE REJECTION

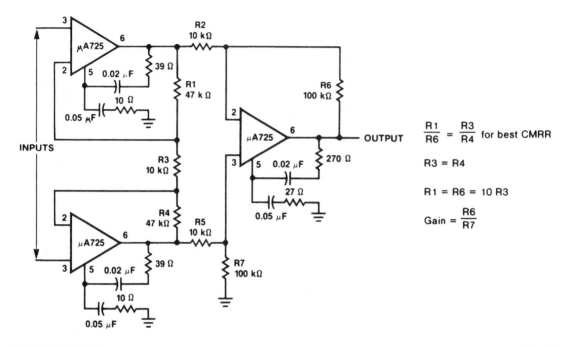

$$\frac{R1}{R6} = \frac{R3}{R4} \text{ for best CMRR}$$

$$R3 = R4$$

$$R1 = R6 = 10\,R3$$

$$\text{Gain} = \frac{R6}{R7}$$

FAIRCHILD CAMERA

Fig. 3-38

LEVEL-SHIFTING ISOLATION AMPLIFIER

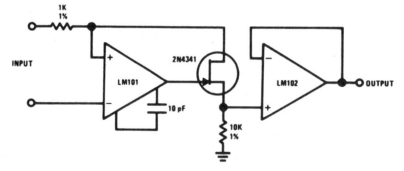

NATIONAL SEMICONDUCTOR

Fig. 3-39

The 2N4341 JFET is used as a level shifter between two op amps operated at different power supply voltages. The JFET is ideally suited for this type of application because $I_D = I_S$.

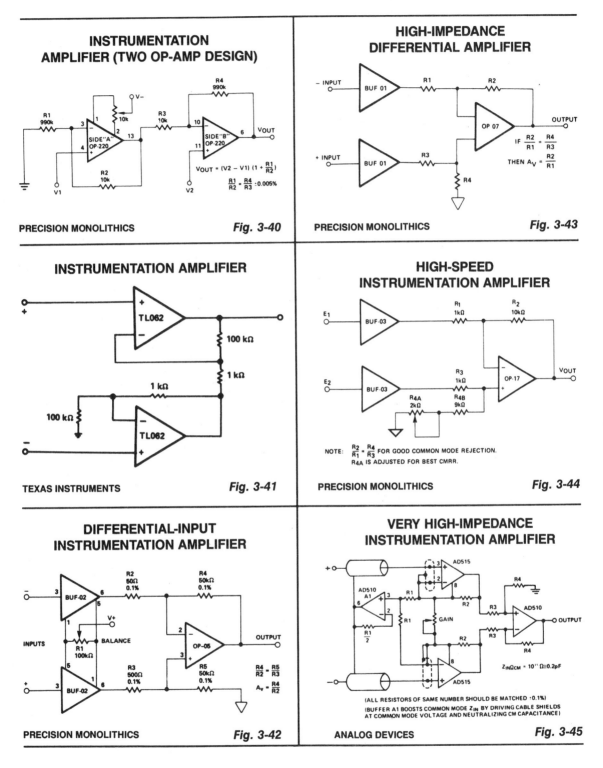

INSTRUMENTATION AMPLIFIER (TWO OP-AMP DESIGN)

R1 990k
R2 10k
R3 10k
R4 990k
V−
10k

SIDE "A" OP-220
3
1
2
13
4
SIDE "B" OP-220
10
6
11

VOUT

V1
V2

$V_{OUT} = (V2 - V1)(1 + \frac{R1}{R2})$

$\frac{R1}{R2} = \frac{R4}{R3} : 0.005\%$

PRECISION MONOLITHICS   *Fig. 3-40*

INSTRUMENTATION AMPLIFIER

+
−
TL062
100 kΩ
1 kΩ
1 kΩ
100 kΩ
−
+
TL062

TEXAS INSTRUMENTS   *Fig. 3-41*

DIFFERENTIAL-INPUT INSTRUMENTATION AMPLIFIER

−
3
BUF-02
6
R2 50Ω 0.1%
R4 50kΩ 0.1%
V+
2
OP-05
OUTPUT
3
BALANCE
R1 100kΩ
INPUTS
5
3
BUF-02
6
R3 500Ω 0.1%
R5 50kΩ 0.1%

$\frac{R4}{R2} = \frac{R5}{R3}$

$A_V = \frac{R4}{R2}$

PRECISION MONOLITHICS   *Fig. 3-42*

HIGH-IMPEDANCE DIFFERENTIAL AMPLIFIER

− INPUT
BUF 01
R1
R2
OP 07
OUTPUT
+ INPUT
BUF 01
R3
R4

IF $\frac{R2}{R1} = \frac{R4}{R3}$

THEN $A_V = \frac{R2}{R1}$

PRECISION MONOLITHICS   *Fig. 3-43*

HIGH-SPEED INSTRUMENTATION AMPLIFIER

E1
BUF-03
R1 1kΩ
R2 10kΩ
OP-17
VOUT
E2
BUF-03
R3 1kΩ
R4A 2kΩ
R4B 9kΩ

NOTE: $\frac{R2}{R1} = \frac{R4}{R3}$ FOR GOOD COMMON MODE REJECTION.
R4A IS ADJUSTED FOR BEST CMRR.

PRECISION MONOLITHICS   *Fig. 3-44*

VERY HIGH-IMPEDANCE INSTRUMENTATION AMPLIFIER

+
AD510 A1
6
3
AD515
2
8
R1
R2
R4
R3
AD510
OUTPUT
$\frac{R1}{2}$
R1
GAIN
R2
R3
R4
AD515

$Z_{IN\Omega CM} = 10^{11} \Omega \| 0.2pF$

(ALL RESISTORS OF SAME NUMBER SHOULD BE MATCHED ±0.1%)
(BUFFER A1 BOOSTS COMMON MODE Z_{IN} BY DRIVING CABLE SHIELDS AT COMMON MODE VOLTAGE AND NEUTRALIZING CM CAPACITANCE)

ANALOG DEVICES   *Fig. 3-45*

AC-COUPLED DYNAMIC AMPLIFIER

This circuit acts as a bandpass filter with gain and would be most useful for biomedical instrumentation. Low-frequency cutoff is set at 10 Hz while the high-frequency breakpoint is given by the open-loop rolloff characteristic of the HA-5141/42/44. In this case, the A_{VCL} = 60 dB where the rolloff occurs at approximately 300 Hz. This corner frequency may be trimmed by inserting a capacitor in parallel with R_f.

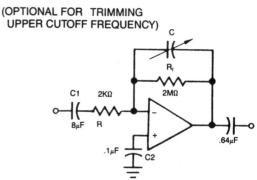

(OPTIONAL FOR TRIMMING UPPER CUTOFF FREQUENCY)

HARRIS   *Fig. 3-46*

FORWARD-CURRENT BOOSTER

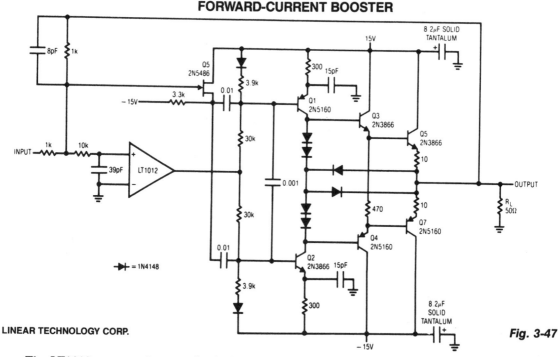

LINEAR TECHNOLOGY CORP.   *Fig. 3-47*

The LT1012 corrects dc errors in the booster stage, and does not set high-frequency signals. Fast signals are fed directly to the stage via Q5 and the 0.01-μF coupling capacitors. Dc and low-frequency signals drive the stage via the op-amp's output. The output stage consists of current sources, Q1 and Q2, driving the Q3 – Q5 and Q4 – Q7 complementary emitter followers. The diode network at the output steers drive away from the transistor bases when output current exceeds 250 mA, providing fast short-circuit protection. The circuit's high frequency summing node is the junction of the 1-K and 10-K resistors at the LT1012. The 10 K/39 pF pair filters high frequencies, permitting accurate dc summation at the LT1012's positive input. This current-boosted amplifier has a slew rate in excess of 1000 V/μs, a full power bandwidth of 7.5 MHz and a 3-dB point of 14 MHz.

RIAA PREAMPLIFIER

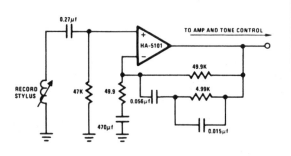

HARRIS Fig. 3-48

The circuit essentially provides low-frequency boost below 318 Hz and high-frequency attenuation above 3150 Hz. Recent modifications to the response standard include a 31.5-Hz peak gain region to reduce dc-oriented distortion from external vibration.

PROFESSIONAL AUDIO
NAB TAPE PLAYBACK PREAMPLIFIER

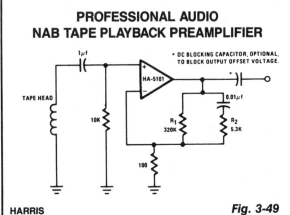

HARRIS Fig. 3-49

The preamplifier is configured to provide low-frequency boost to 50 Hz, flat response to 3 kHz, and high-frequency attenuation above 3 kHz. Compensation for variations in tape and tape head performance can be achieved by trimming R1 and R2.

AUTOMATIC LEVEL CONTROL

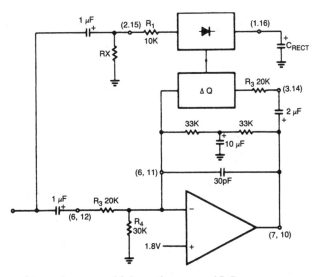

SIGNETICS Fig. 3-50

The NE570 can be used to make a very high-performance ALC compressor, except that the rectifier input is tied to the input. This makes gain inversely proportional to input level so that a 20-dB drop in input level will produce a 20 dB increase in gain. The output will remain fixed at a constant level. As shown, the circuit will maintain an output level of ±1 dB for an input range of +14 to −43 dB at 1 kHz. Additional external components will allow the output level to be adjusted.

× 1000 AMPLIFIER

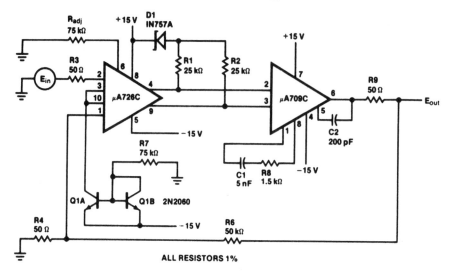

FAIRCHILD CAMERA

Fig. 3-51

TWO-WIRE TO FOUR-WIRE AUDIO CONVERTER

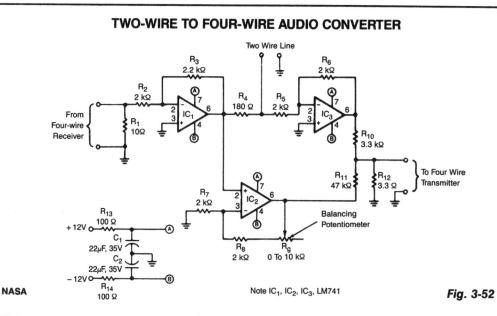

NASA

Note IC_1, IC_2, IC_3, LM741

Fig. 3-52

This converter circuit maintains 40 dB of isolation between the input and output halves of a four-wire line while permitting a two-wire line to be connected. A balancing potentiometer, Rg, adjusts the gain of IC2 to null the feed-through from the input to the output. The adjustment is done on the workbench just after assembly by inserting a 1-kHz tone into the four-wire input and setting R_g for minimum output signal. An 82-Ω dummy-load resistor is placed across the two wire terminals.

LOW-POWER INSTRUMENTATION AMPLIFIER

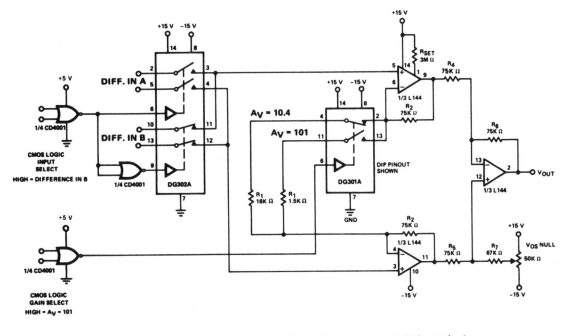

RSET programs L144 power dissipation, gain-bandwidth product. Refer to AN73-6 and the L144 data sheet.

Voltage gain of the instrumentation amplifier is:

$$A_V = 1 + \frac{2R_2}{R_1} \quad \text{(In the circuit shown, } A_{V1} = 10.4, A_{V2} = 101)$$

SILICONIX

Fig. 3-53

4

Logic Amplifiers

The sources of the following circuits are contained in the Sources section, which begins on page 182. The figure number contained in the box of each circuit correlates to the source entry in the Sources section.

Low-Power Inverting Amplifier with Digitally Selectable Gain
Low-Power Low-Frequency Amplifier
Low-Power Noninverting Amplifier with Digitally Selectable Inputs and Gain
Programmable Amplifier
Precision Amplifier with Digitally Programmable Inputs and Gain

LOW-POWER INVERTING AMPLIFIER WITH DIGITALLY SELECTABLE GAIN

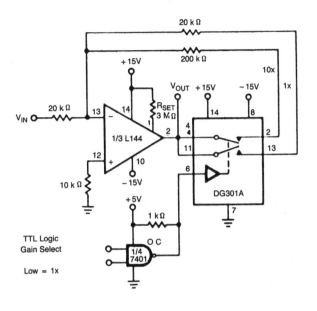

SILICONIX

Fig. 4-1

LOW-POWER LOW-FREQUENCY AMPLIFIER

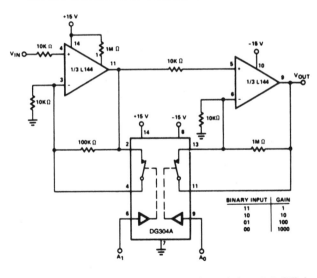

SILICONIX

Fig. 4-2

Gain increases by decades as the binary input decreases from 1,1 to 0,0. Minimum gain is 1 and maximum gain is 1000. Because the switch is static in this type of amplifier the power dissipation of the switch will be less than a tenth of a milliwatt.

LOW-POWER NONINVERTING AMPLIFIER WITH DIGITALLY SELECTABLE INPUTS AND GAIN

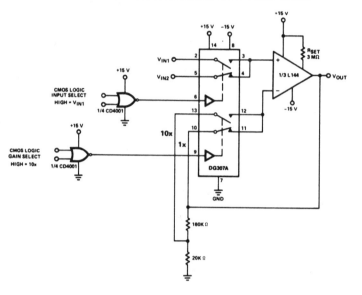

SILICONIX

Fig. 4-3

PROGRAMMABLE AMPLIFIER

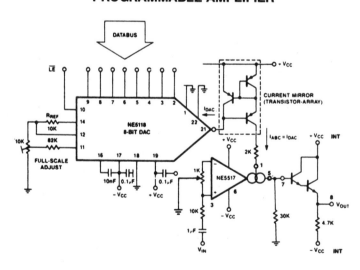

SIGNETICS

Fig. 4-4

The intention of the following application shows how the NE5517 works in connection with a DAC. In the application, the NE5118 is used—an 8-bit DAC with current output—its input register making this device fully μP-compatible. The circuit consists of three functional blocks; the NE5118 which generates a control current equivalent to the applied data byte, a current mirror, and the NE5517.

PRECISION AMPLIFIER WITH DIGITALLY
PROGRAMMABLE INPUTS AND GAIN

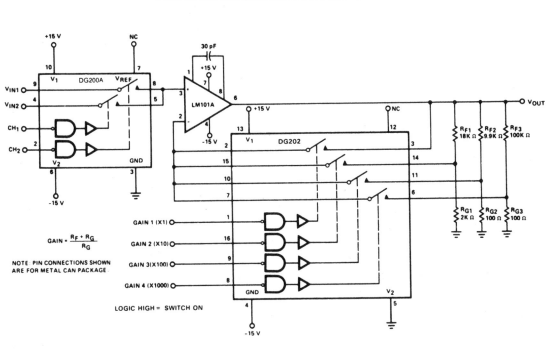

$$GAIN = \frac{R_F + R_G}{R_G}$$

NOTE: PIN CONNECTIONS SHOWN
ARE FOR METAL CAN PACKAGE.

LOGIC HIGH = SWITCH ON

5

Mixers, Crossovers and Distribution Amplifiers

The sources of the following circuits are contained in the Sources section, which begins on page 182. The figure number contained with each circuit correlates to the source entry in the Sources section.

ELECTRONIC CROSSOVER CIRCUIT

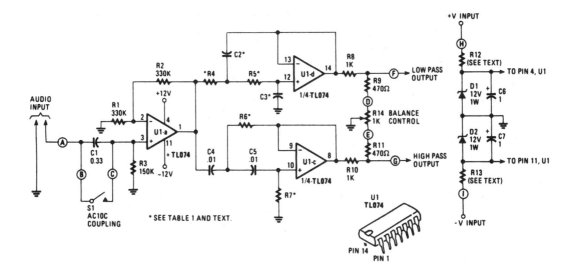

Fig. 5-1

An audio source (such as a mixer, preamplifier, equalizer, or recorder) is fed to the Electronic Cross-over Circuit's input. That signal is either ac- or dc-coupled, depending on the setting of switch S1, to the noninverting input of buffer-amplifier U1a, one section of a quad, BIFET, low-noise TL074 op amp made by Texas Instruments. That stage has a gain of 2, and its output is distributed to both a lowpass filter made by R4, R5, C2, C3, and op-amp U1d, and a highpass filter made by R6, R7, C4, C5, and op amp U1c. Those are 12-dB/octave Butterworth-type filters. The Butterworth filter response was chosen because it gives the best compromise between damping and phase shift. Values of capacitors and resistors will vary with the selected crossover at which your unit will operate. The filter's outputs are fed to a balancing net-work made by R8, R9, R10, R11 and balance potentiometer R14. When the potentiometer is at its mid-position, there is unity gain for the passbands of both the high and low filters. Dc power for the Electronic Crossover Circuit is regulated by R12, R13, D1, and D2, and decoupled by C6 and C7.

SOUND MIXER/AMPLIFIER

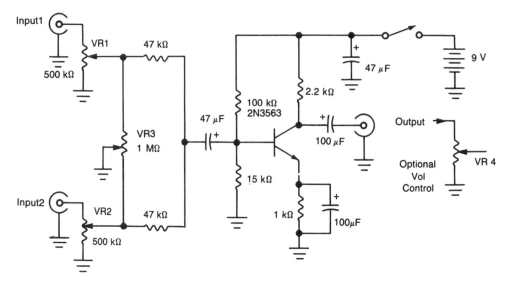

WILLIAM SHEETS

Fig. 5-2

Both input signals can be independently controlled by VR1 and VR2. The balance control VR3 is used to fade out one signal while simultaneously fading in the other. The transistor provides gain and the combined output signal level is controlled by VR4 (optional).

MICROPHONE MIXER

A TL081 op amp is used as a high-to-low imped-ance converter and signal mixer. The input imped-ance is approximately 1 megohm and the output impedance is about 1 kΩ. Two 9-V batteries are used as the power source. Battery life should be several hundred hours with alkaline batteries.

WILLIAM SHEETS

Fig. 5-3

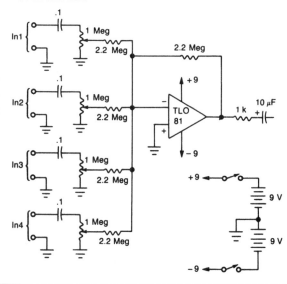

LOW-DISTORTION INPUT SELECTOR

LOW DISTORTION INPUT SELECTOR

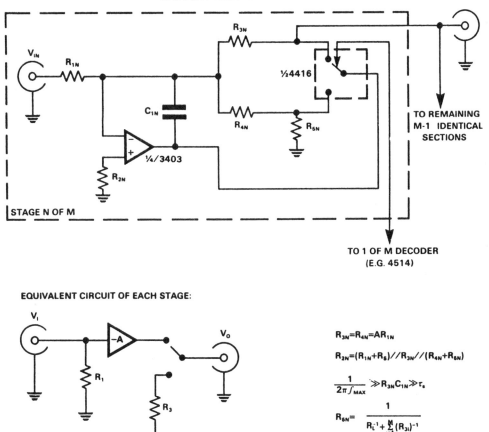

EQUIVALENT CIRCUIT OF EACH STAGE:

$R_{3N}=R_{4N}=AR_{1N}$

$R_{2N}=(R_{1N}+R_6)//R_{3N}//(R_{4N}+R_{6N})$

$\dfrac{1}{2\pi f_{MAX}} \gg R_{3N}C_{1N} \gg r_s$

$R_{6N}= \dfrac{1}{R_L^{-1}+\sum\limits_{i=1}^{M}(R_{3i})^{-1}}$

ELECTRONIC ENGINEERING

Fig. 5-4

CMOS switches are used directly to select inputs in audio circuits, this can introduce unacceptable levels of distortion, but if the switch is included in the feedback network of an op amp, the distortion as a result to the switch can be almost eliminated. The circuit uses a 4416 CMOS switch, arranged as two independent SPDT switches. If switching transients are unimportant, R5 and C1 can be omitted, and R4 can be shorted out. However, a feedback path must be maintained, even when a channel is switched out, in order to keep the inverting input of the op amp at ground potential, and prevent excessive crosstalk between channels.

AUDIO DISTRIBUTION AMPLIFIER

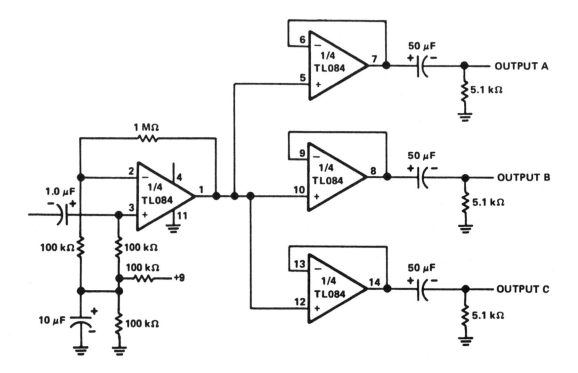

TEXAS INSTRUMENTS

Fig. 5-5

The 3-channel output distribution amplifier uses a single TL084. The first stage is capacitively coupled with a 1.0-μF electrolytic capacitor. The inputs are at $1/2$-V_{CC} rail or 4.5 V. This makes it possible to use a single 9-V supply. A voltage gain of 10 (1-MΩ/100-kΩ) is obtained in the first stage, and the other three stages are connected as unity-gain voltage followers. Each output stage independently drives an amplifier through the 50-μF output capacitor to the 5.1-kΩ load resistor. The response is flat from 10 Hz to 30 kHz.

SIGNAL DISTRIBUTION AMPLIFIER

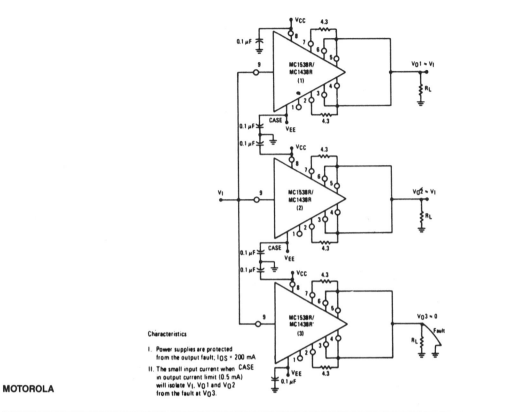

Characteristics

I. Power supplies are protected from the output fault; $I_{OS} = 200$ mA

II. The small input current when in output current limit (0.5 mA) will isolate V_I, V_{O1} and V_{O2} from the fault at V_{O3}.

MOTOROLA

Fig. 5-6

AUDIO DISTRIBUTION AMPLIFIER

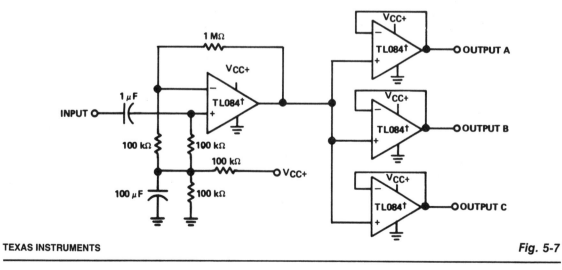

TEXAS INSTRUMENTS

Fig. 5-7

FOUR-CHANNEL FOUR-TRACK MIXER

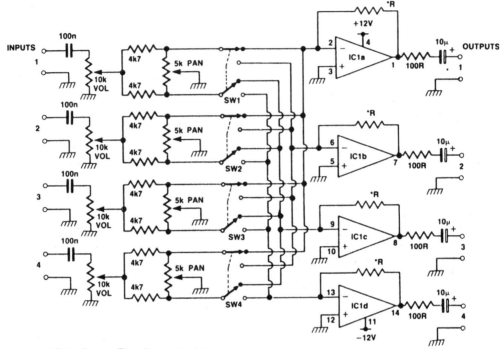

*Note: Choose R to give appropriate gain.

Fig. 5-8

This circuit can be used as a stereo mixer as well as a four-track mixer. The quad op-amp IC gives a bit of gain for each track. The pan control allows panning between tracks 1 and 2 with the switch in the up position; with the switch in the down position, it is possible to pan between tracks 3 and 4. Extra channels can be added. A suitable op amp for IC1 is TL074 or similar.

6

Operational Amplifiers

The sources of the following circuits are contained in the Sources section, which begins on page 182. The figure number contained in the box of each circuit correlates to the source entry in the Sources section.

continued

Gated Amplifier
Reference-Voltage Amplifier
Fast Summing Amplifier
Adjustment-Free Precision Summing Amplifier
Color Video Amplifier
Fast Voltage Follower
Isolation Amplifier for Capacitive Loads
Cable Bootstrapping Shield
Logarithmic Amplifier
Voltage-Controlled Variable-Gain Amplifier
Inverting Amplifier with Balancing
Switching Power Amplifier
Precision Power Booster
Noninverting Voltage Follower
High-Impedance Differential Amplifier
Unity-Gain Follower
Variable-Gain and Sign Op Amp
Discrete Current Booster

Precision Process-Control Interface
Intrinsically Safe Protected Op Amp
± 15-V Chopper Amplifier
Input/Output Buffer Amplifier for Analog
   Multiplexers
Absolute-Value Norton Amplifier
High-Input-Impedance Differential Amplifier
Hi-Fi Compandor
Active-Clamp Limiting Amplifier
Wide-Band AGC Amplifier
Audio Automatic Gain Control
Chopper-Stabilized Amplifier
Ultra-Low-Leakage Preamplifier
FET Input Amplifier
Ultra-High Z_{in} ac Unity-Gain Amplifier
Stereo Amplifier with Gain Control
Noninverting Amplifier

OPERATIONAL AMPLIFIERS

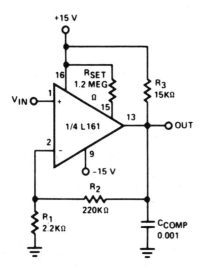

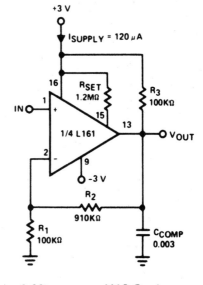

The L161 as a X100 Operational Amplifier

(A)

A Micropower X10 Op Amp

(B)

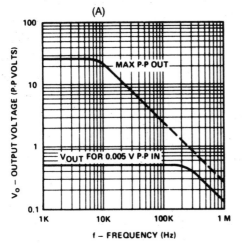

**Frequency Response and Maximum Output
for the X100 Op Amp**

SILICONIX

Fig. 6-1

This is a single gain-of-100 amplifier with a gain-bandwidth product of 20 MHz! The primary limitation in the performance is the low slew rate (0.3 V/μs) imposed by I_{OH} charging C_{COMP}. The effects of slew rate and compensation are shown. A lower gain amplifier requires a larger C_{COMP}, which in turn further reduces slew rate. For this reason, it might actually be advantageous in certain areas to lower the gain by placing a resistive divider at the input rather than raising R_I. Figure 6-1B shows a 700-μW, $\times 10$ op amp whose slew rate is 0.02 V/μs and is 3 dB down at 100 kHz.

VOLTAGE-FOLLOWER AMPLIFIER FOR SIGNAL-SUPPLY OPERATION

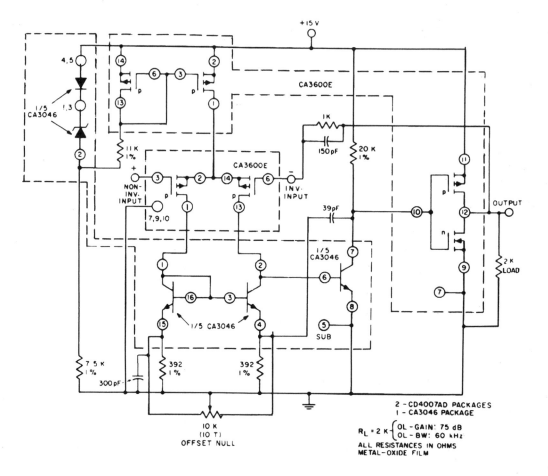

INTERSIL

Fig. 6-2

This unity-gain follower amplifier has a CMOS p-channel input, an npn second-gain stage, and a CMOS inverter output. The IC building blocks are two CA3600E's (CMOS transistor pairs) and a CA3046 npn transistor array. A zener-regulated leg provides bias for a 400-μA p-channel source, feeding the input stage, which is terminated in an npn current mirror. The amplifier voltage-offset is nulled with the 10-KΩ balance potentiometer. The second-stage current level is established by the 20-KΩ load, and is selected to approximately the first-stage current level, to assure similar positive and negative slew rates. The CMOS inverter portion forms the final output stage and is terminated in a 2-KΩ load, a typical value used with monolithic op amps. Voltage gain is affected by the choice of load resistance value. The output stage of this amplifier is easily driven to within 1 mV of the negative supply voltage.

OP-AMP GAIN ADJUSTER

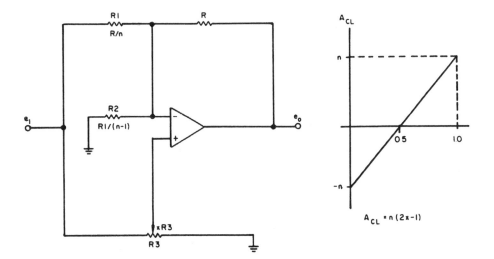

ELECTRONIC DESIGN

Fig. 6-3

An op amp's gain level can be adjusted over its full inverting and noninverting gain range. R3 varies the signal applied to both the inverting and noninverting amplifier inputs. When the wiper position (denoted by x) equals zero, the noninverting amplifier input is grounded. This also holds the voltage across R2 at zero, so R2 has no effect on operation. Now only R1 and R carry feedback current, and the amplifier operates at a gain of $-n$. At the other pot extreme, where $x = 1$, the input signal is connected directly to the noninverting input. Because feedback maintains a near-zero voltage between the amplifier inputs, the amplifier's inverting input will also be near the input signal level, thus little voltage is across R1. Also, the gain is now $+n$. The amplifier should be driven from a low-impedance source to minimize source-loading error and low-offset op amps should be used.

LOG-RATIO AMPLIFIER

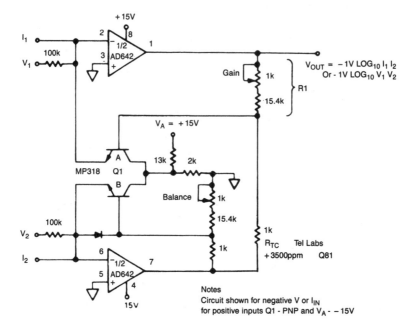

$V_{OUT} = -1V \ LOG_{10} \ I_1 \ I_2$
Or $-1V \ LOG_{10} \ V_1 \ V_2$

Notes
Circuit shown for negative V or I_{IN}
for positive inputs Q1 - PNP and V_A - $-15V$

ANALOG DEVICES

Fig. 6-4

INVERTING AMPLIFIER

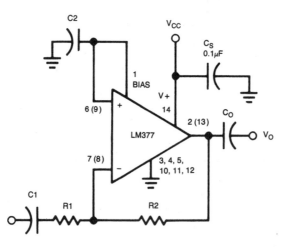

NATIONAL SEMICONDUCTOR

Fig. 6-5

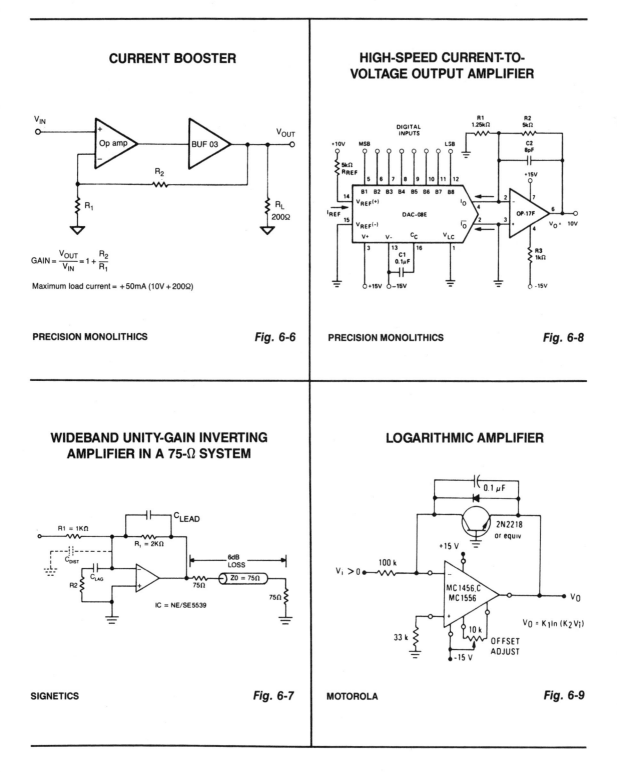

CURRENT BOOSTER

$$\text{GAIN} = \frac{V_{OUT}}{V_{IN}} = 1 + \frac{R_2}{R_1}$$

Maximum load current = +50mA (10V + 200Ω)

PRECISION MONOLITHICS

Fig. 6-6

HIGH-SPEED CURRENT-TO-VOLTAGE OUTPUT AMPLIFIER

PRECISION MONOLITHICS

Fig. 6-8

WIDEBAND UNITY-GAIN INVERTING AMPLIFIER IN A 75-Ω SYSTEM

SIGNETICS

Fig. 6-7

LOGARITHMIC AMPLIFIER

$V_O = K_1 \ln (K_2 V_1)$

MOTOROLA

Fig. 6-9

VOLTAGE-CONTROLLED AMPLIFIER

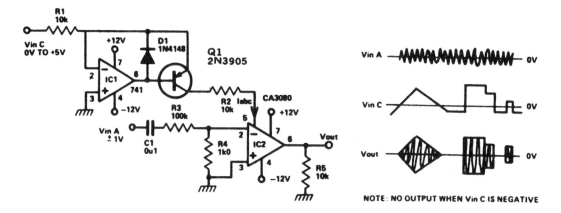

ELECTRONICS TODAY INTERNATIONAL

Fig. 6-10

This circuit is basically an op amp with an extra input at pin 5. Current I_{ABC} is injected into this input and this controls the gain of the device linearly. By inserting an audio signal (± 10 mV) between pin 2 and 3 and by controlling the current on pin 5, the level of the signal output (pin 6) is controlled.

ABSOLUTE-VALUE AMPLIFIER

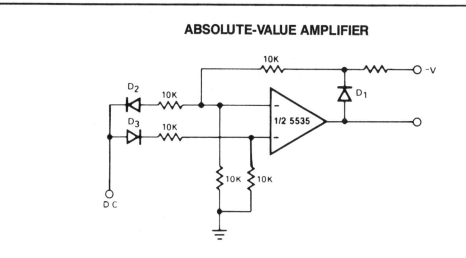

SIGNETICS

Fig. 6-11

The circuit generates a positive output voltage for either polarity of input. For positive signals, it acts as a noninverting amplifier and for negative signals, it acts as an inverting amplifier. The accuracy is poor for input voltages under 1 V, but for less-stringent applications, it can be effective.

CURRENT-SHUNT AMPLIFIER

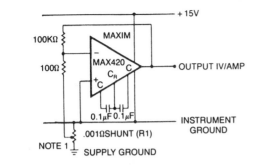

This circuit measures the power-supply current of a circuit without really having a current-shunt resistor: R1 is only 3 cm of #20 gauge copper wire. A length of the power distribution wiring can be used for R1. The MAX420's CMVR includes its own negative power supply; therefore, it can both be powered by and measure current in the ground line.

MAXIM

Fig. 6-12

CONSTANT-BANDWIDTH AMPLIFIER

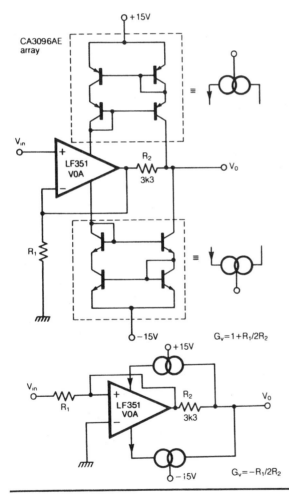

The traditional restriction of constant gain-bandwidth products for a voltage amplifier can be overcome by employing feedback around a current amplifier. Two current mirrors, constructed from transistors in a CA3096AE array, effectively turn the LF351 op amp into a current amplifier. Feedback is then applied by using R2 and R1, turning the whole circuit into a feedback voltage amplifier with a noninverting gain of G of $1 + R_1/2R_2$.

Using the values shown, a constant bandwidth of 3.5 MHz is obtained for all voltage gains up to and beyond 100 at 10 V pk-pk output, equivalent to a gain-bandwidth product of 350 MHz from an op amp with an advertised unity gain-bandwidth of 10 MHz. An inverting gain configuration is also possible (see Fig. 2) where $G_v = -R_1/2R_2$. Slewing rates are significantly improved by this approach; even a 741 can manage 100 V μs under these conditions since its output is a virtual earth. However, because the new configurations use current feedback to achieve bandwidth independence, an output buffer should be added for circuits where a significant output current is required.

ELECTRONIC ENGINEERING

Fig. 6-13

93

POLARITY-REVERSING LOW-POWER AMPLIFIER

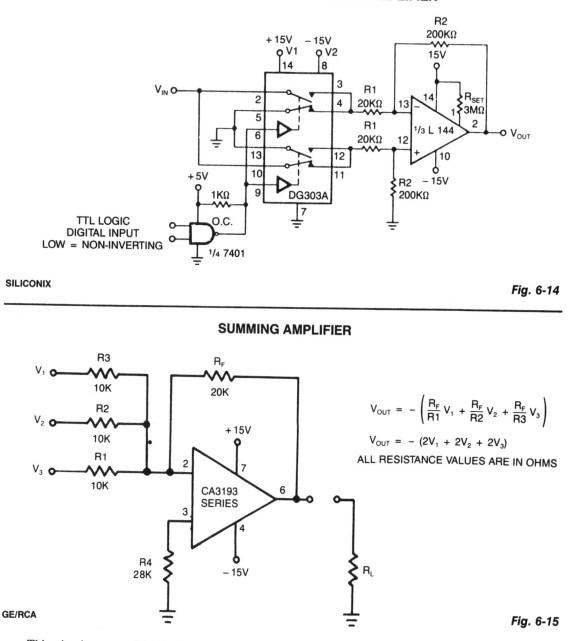

SILICONIX

Fig. 6-14

SUMMING AMPLIFIER

$$V_{OUT} = - \left(\frac{R_F}{R1} V_1 + \frac{R_F}{R2} V_2 + \frac{R_F}{R3} V_3 \right)$$

$$V_{OUT} = - (2V_1 + 2V_2 + 2V_3)$$

ALL RESISTANCE VALUES ARE IN OHMS

GE/RCA

Fig. 6-15

This circuit uses a CA3193 BiMOS op amp. Because input noise of the amplifier is increased by R_F/R1//R2//R3, and the gain that a single input will amplify is the gain of only one of the input channels (R_F/R1), for good noise performance, use the smallest number of inputs.

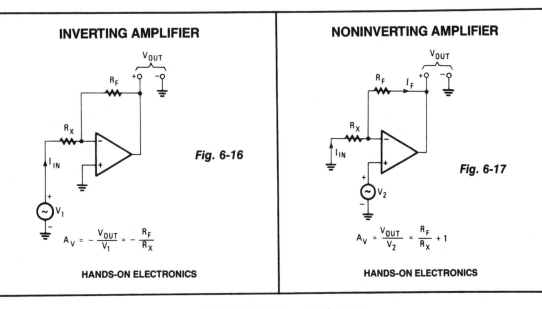

INVERTING AMPLIFIER

Fig. 6-16

$$A_V = -\frac{V_{OUT}}{V_1} = -\frac{R_F}{R_X}$$

HANDS-ON ELECTRONICS

NONINVERTING AMPLIFIER

Fig. 6-17

$$A_V = \frac{V_{OUT}}{V_2} = \frac{R_F}{R_X} + 1$$

HANDS-ON ELECTRONICS

DIFFERENTIAL AMPLIFIER

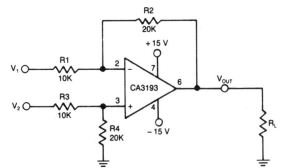

ALL RESISTANCE VALUES ARE IN OHMS

$$V_{OUT} = V_2 \left(\frac{R4}{R3 + R4} \right) \left(\frac{R1 + R2}{R1} \right) - V_1 \left(\frac{R2}{R1} \right)$$

IF R4 = R2, R3 = R1 AND $\frac{R2}{R1} = \frac{R4}{R3}$

THEN $V_{OUT} = (V_2 - V_1) \left(\frac{R2}{R1} \right)$

FOR VALUES ABOVE $V_{OUT} = (V_2 - V_1)$

IF A_V IS TO BE MADE 1 AND IF R1 = R3 = R4 = R
WITH R2 = 0.999R (0.1% MISMATCH IN R2)

THEN $V_{OCM} = 0.0005\ V_{IN}$ OR CMRR = 66 dB
THUS, THE CMRR OF THIS CIRCUIT IS LIMITED BY
THE MATCHING OR MISMATCHING OF THIS NETWORK
RATHER THAN THE AMPLIFER

This differential amplifier uses a CA3193 BiMOS op amp. This classical, differential input-to-signal-ended output converter when used with low-resistance signal source will maintain level of CMRR, if $R_1 = R_3 + R_4$.

GE/RCA

Fig. 6-18

COMPOSITE AMPLIFIER

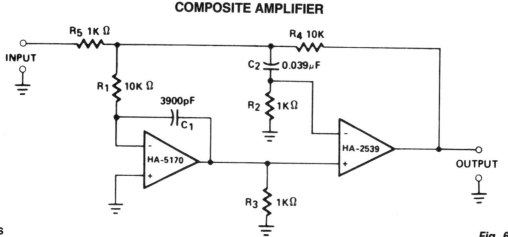

R5 1K Ω

INPUT

R1 10K Ω

3900pF C1

HA-5170

C2 0.039µF

R4 10K

R2 1KΩ

HA-2539

OUTPUT

R3 1KΩ

HARRIS

Fig. 6-19

A composite configuration greatly reduces dc errors without compromising the high-speed, wideband characteristics of HA-2539. The HA-2540 could also be used, but with slightly lower speeds and bandwidth response.

The HA-2539 amplifies signals above 40 kHz which are fed forward via C2; R2 and R5 set the voltage gain at −10. The slew rate of this circuit was measured at 350 V/μs. Settling time to a 0.1% level for a 10-V output step is under 150 ns and the gain bandwidth product is 300 MHz.

The HA-5170 amplifies signals below 40 kHz, as set by C1 and R1, and controls the dc input characteristics such as offset voltage, drift, and bias currents of the composite amplifier. Therefore, it has an offset voltage of 100 μV, drift of 2 μV/°C, and bias currents in the 20-pA range. The offset voltage can be externally nulled by connecting a 20-KΩ pot to pins 1 and 5, with the wiper tied to the negative supply. The dc gains of the HA-5170 and HA-2539 are cascaded; this means that the dc gain of the composite amplifier is well over 160 dB.

The excellent ac and dc performance of this composite amplifier is complemented by its low noise performance, 0.5-μV rms from 0.1 Hz to 100 Hz. It is very useful in high-speed data acquisition systems.

CASCADED AMPLIFIER

Cascaded amplifier sections extend bandwidth and increase gain. Using two HA-2539 devices, this circuit is capable of 60-dB gain at 20 MHz.

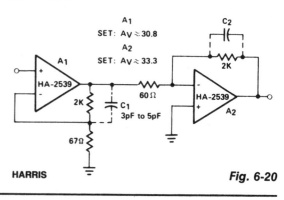

A1
SET: Av ≈ 30.8

A2
SET: Av ≈ 33.3

A1
HA-2539

2K

C1
3pF to 5pF

67Ω

60Ω

C2

2K

HA-2539
A2

HARRIS

Fig. 6-20

96

PULSE-WIDTH PROPORTIONAL-CONTROLLER CIRCUIT

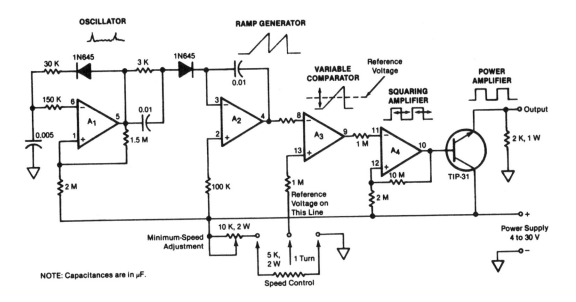

NASA

Fig. 6-21

The quad operational-amplifier circuit yields full 0 to 100 percent pulse-width control. The controller uses an LM3900 that requires only a single supply voltage of 4 to 30 V. The pulse-repetition rate is set by a 1-kHz oscillator that incorporates amplifier A_1. The oscillator feeds ramp generator A_2, which generates a linear ramp voltage for each oscillator pulse. The ramp signal feeds the inverting input of comparator A_3; the speed-control voltage feeds the noninverting input. Thus, the output of the comparator is a 1-kHz pulse train, the pulse width of which changes linearly with the control voltage. The control voltage can be provided by an adjustable potentiometer or by an external source of feedback information such as a motor-speed sensing circuit. Depending on the control-voltage setting, the pulse duration can be set at any value from zero (for zero average dc voltage applied to the motor) to the full pulse-repetition period (for applied motor voltage equal to dc power-supply voltage). An amplifier stage (A_4) with a gain of 10 acts as a pulse-squaring circuit. A TIP-31 medium-power transistor is driven by A_4 and serves as a separate power-amplifier stage.

OP-AMP CLAMPER

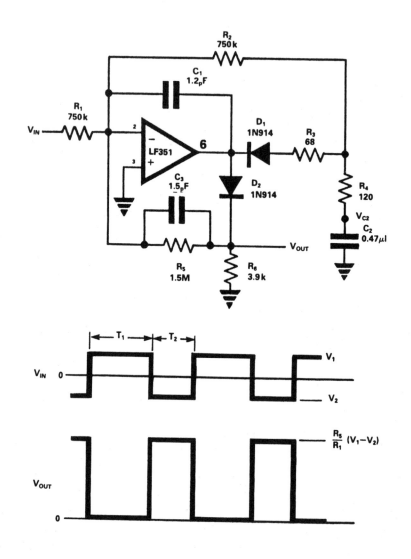

ELECTRONIC ENGINEERING

Fig. 6-22

The circuit clamps the most positive value of the input pulse signal to the zero base level. Additionally, the circuit inverts and amplifies the input signal by the factor of R_5/R_1. The waveforms are shown in the bottom of Fig. 6-22.

VOLTAGE-CONTROLLED ATTENUATOR

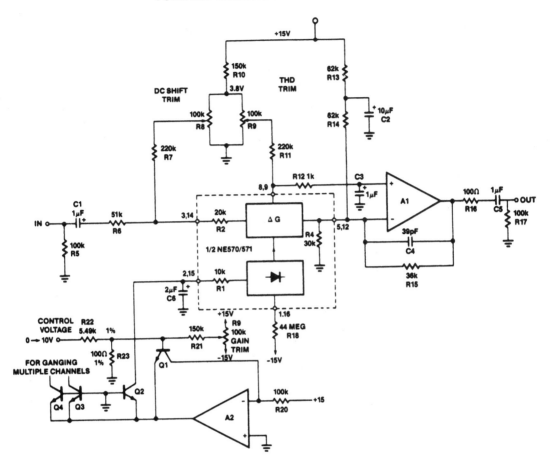

SIGNETICS

Fig. 6-23

Op amp A_2 and transistors Q_1 and Q_2 form the exponential converter generating an exponential gain-control current, which is fed into the rectifier. A reference current of 150 μA, (15 V and R_{20} = 100 kΩ), is attenuated a factor of two (6 dB) for every volt increase in the control voltage. Capacitor C6 slows down gain changes to a 20-ms time constant ($C_6 \times R_1$) so that an abrupt change in the control voltage will produce a smooth sounding gain change. R18 ensures that for large control voltages the circuit will go to full attenuation. The rectifier bias current would normally limit the gain reduction to about 70 dB. R16 draws excess current out of the rectifier. After approximately 50 dB of attenuation at a −6-dB/V slope, the slope steepens and attenuation becomes much more rapid until the circuit totally shuts off at about 9 V of control voltage. A1 should be a low-noise high slew rate op amp. R13 and R14 establish approximately a 0-V bias at A1's output.

LOGARITHMIC AMPLIFIER

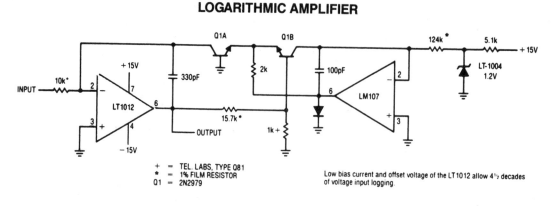

+ = TEL. LABS, TYPE Q81
* = 1% FILM RESISTOR
Q1 = 2N2979

Low bias current and offset voltage of the LT1012 allow 4½ decades
of voltage input logging.

Fig. 6-24

COMPOSITE AMPLIFIER

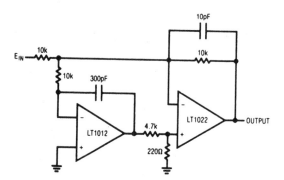

Fig. 6-25

The circuit is made up of an LT1012 low-drift device, and an LT1022 high-speed amplifier. The overall circuit is a unity-gain inverter, and the summing node is located at the junction of three 10-kΩ resistors. The LT1012 monitors this summing node, compares it to ground, and drives the LT1022's positive input, completing a dc stabilizing loop around the LT1022. The 10-kΩ/300-pF time constant at the LT1012 limits its response to low-frequency signals. The LT1022 handles high-frequency inputs while the LT1012 stabilizes the dc operating point. The 4.7-kΩ/220-Ω divider at the LT1022 prevents excessive input overdrive during start-up. This circuit combines the LT1012's 35-μV offset and 1.5-V/°C drift with the LT1022's 23-V/μs slew rate and 300-kHz full-power bandwidth. Bias current, dominated by the LT1012, is about 100 pA.

STABLE UNITY-GAIN BUFFER

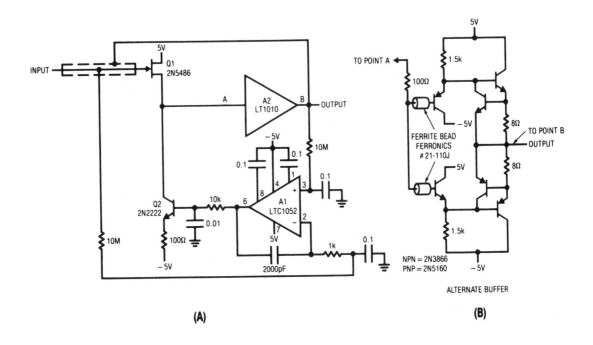

(A)   (B)

LINEAR TECHNOLOGY CORP.   *Fig. 6-26*

Q1 and Q2 constitute a simple, high-speed FET input buffer. Q1 functions as a source follower, with the Q2 current source load setting the drain-source channel current. Normally, this open-loop configuration would be quite drifty because it has no dc feedback. The LTC1052 contributes this function to stabilize the circuit by comparing the filtered circuit output to a similarly filtered version of the input signal. The amplified difference between these signals is used to set Q2's bias, and hence Q1's channel current. This forces Q1's V_{GS} to whatever voltage is required to match the circuit's input and output potentials. The 2000-pF capacitor at A1 provides stable loop compensation. The RC network in A1's output prevents it from seeing high-speed edges that are coupled through Q2's collector-base junction. A2's output is also fed back to the shield around Q1's gate lead, which bootstraps the circuit's effective input capacitance down to less than 1 pF. For very fast requirements, the alternate discrete component buffer shown will be useful. Although its output is current limited at 75 mA, the GHz-range transistors employed provide exceptionally wide bandwidth, fast slewing, and very little delay.

HIGH INPUT IMPEDANCE HIGH-OUTPUT CURRENT-VOLTAGE FOLLOWER

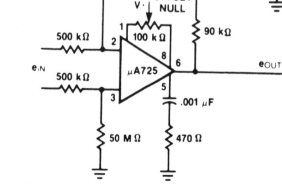

V_{CC}

0.1 μF

MC1456,C
MC1556

MC1438R
MC1538R

V_I

$z_i = 250\ M\Omega$

10 k

OFFSET
ADJUST

0.1 μF

CASE

1 k

470 pF

V_o

$z_0 = 100\ \mu\Omega$
$I_0 = 100\ mA\ (max)$

V_{EE}

MOTOROLA

Fig. 6-27

PRECISION AMPLIFIER

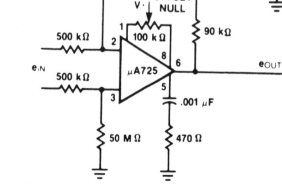

50 MΩ

10 k Ω

OFFSET
NULL

V·I

500 kΩ

100 kΩ

90 kΩ

e_{IN}

500 kΩ

μA725

e_{OUT}

.001 μF

50 MΩ

470 Ω

Characteristics
$A_V = 1000 = 60\ dB$
DC Gain Error = 0.05%
Bandwidth = 1 kHz for −0.05% error
Diff. Input Res. = 1 MΩ
Typical amplifying capability
$e_{IN} = 10\ \mu V$ on $V_{CMI} = 1.0\ V$
Caution: Minimize Stray Capacitance
$A_{VCL} = 1000$

Pin numbers are shown for metal package only

FAIRCHILD CAMERA

Fig. 6-28

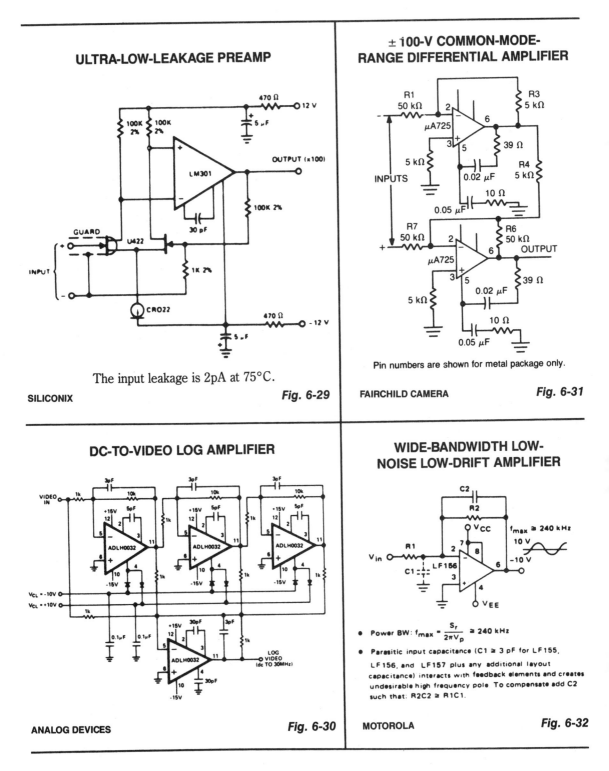

ULTRA-LOW-LEAKAGE PREAMP

The input leakage is 2pA at 75°C.

SILICONIX   *Fig. 6-29*

± 100-V COMMON-MODE-RANGE DIFFERENTIAL AMPLIFIER

Pin numbers are shown for metal package only.

FAIRCHILD CAMERA   *Fig. 6-31*

DC-TO-VIDEO LOG AMPLIFIER

ANALOG DEVICES   *Fig. 6-30*

WIDE-BANDWIDTH LOW-NOISE LOW-DRIFT AMPLIFIER

* Power BW: $f_{max} = \dfrac{S_r}{2\pi V_p} \geq 240$ kHz

* Parasitic input capacitance (C1 ≥ 3 pF for LF155, LF156, and LF157 plus any additional layout capacitance) interacts with feedback elements and creates undesirable high frequency pole. To compensate add C2 such that: R2C2 ≥ R1C1.

MOTOROLA   *Fig. 6-32*

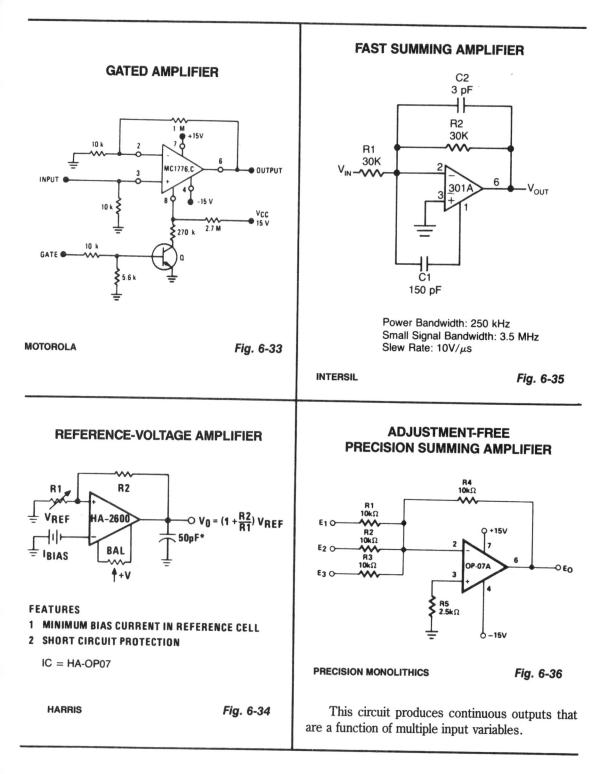

GATED AMPLIFIER

INPUT

OUTPUT

GATE

MC1776.C

MOTOROLA   *Fig. 6-33*

FAST SUMMING AMPLIFIER

C2
3 pF

R2
30K

R1
30K

V_{IN}

301A

V_{OUT}

C1
150 pF

Power Bandwidth: 250 kHz
Small Signal Bandwidth: 3.5 MHz
Slew Rate: 10V/μs

INTERSIL   *Fig. 6-35*

REFERENCE-VOLTAGE AMPLIFIER

R1   R2

V_{REF} HA-2600

I_{BIAS} BAL

$V_0 = (1 + \frac{R2}{R1}) V_{REF}$

50pF*

+V

FEATURES
1 MINIMUM BIAS CURRENT IN REFERENCE CELL
2 SHORT CIRCUIT PROTECTION

IC = HA-OP07

HARRIS   *Fig. 6-34*

ADJUSTMENT-FREE
PRECISION SUMMING AMPLIFIER

R4
10kΩ

R1
10kΩ

E_1

R2
10kΩ

E_2

R3
10kΩ

E_3

OP-07A

+15V

E_O

−15V

R5
2.5kΩ

PRECISION MONOLITHICS   *Fig. 6-36*

This circuit produces continuous outputs that are a function of multiple input variables.

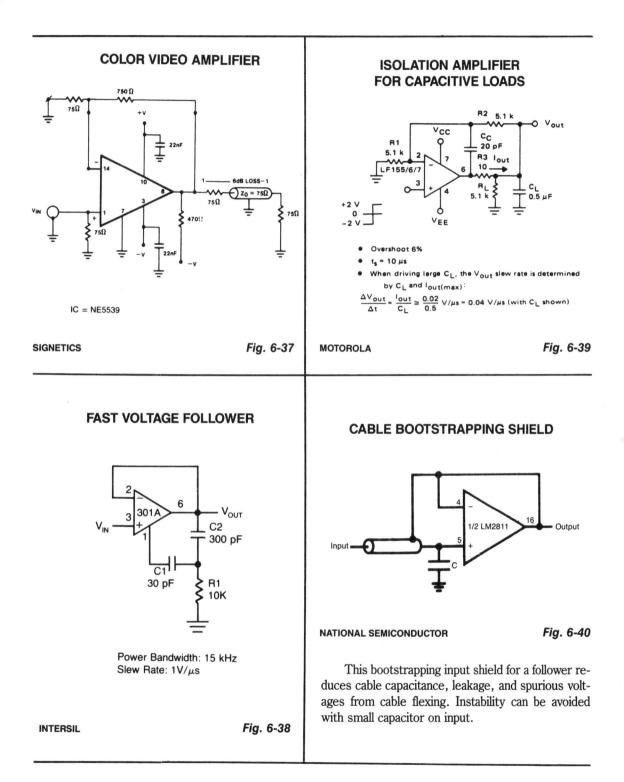

COLOR VIDEO AMPLIFIER

750 Ω

75 Ω

+V

22nF

−

14

10

8

3

1

7

6dB LOSS−1

$Z_0 = 75Ω$

75Ω

75Ω

V_{IN}

+

75Ω

470!!

−V

22nF

−V

IC = NE5539

SIGNETICS   *Fig. 6-37*

ISOLATION AMPLIFIER
FOR CAPACITIVE LOADS

V_{CC}

R1
5.1 k

R2 5.1 k

C_C
20 pF

V_{out}

2

−

7

R3 I_{out}
10

LF155/6/7

3

+

6

4

R_L
5.1 k

C_L
0.5 µF

+2 V
0
−2 V

V_{EE}

- Overshoot 6%
- $t_s = 10 \mu s$
- When driving large C_L, the V_{out} slew rate is determined by C_L and $I_{out(max)}$:

$$\frac{\Delta V_{out}}{\Delta t} = \frac{I_{out}}{C_L} \cong \frac{0.02}{0.5} \text{ V}/\mu s = 0.04 \text{ V}/\mu s \text{ (with } C_L \text{ shown)}$$

MOTOROLA   *Fig. 6-39*

FAST VOLTAGE FOLLOWER

2

−

6

V_{OUT}

3

301A

+

1

V_{IN}

C2
300 pF

C1
30 pF

R1
10K

Power Bandwidth: 15 kHz
Slew Rate: 1V/μs

INTERSIL   *Fig. 6-38*

CABLE BOOTSTRAPPING SHIELD

4

−

16

1/2 LM2811

Output

Input

5

+

C

NATIONAL SEMICONDUCTOR   *Fig. 6-40*

This bootstrapping input shield for a follower reduces cable capacitance, leakage, and spurious voltages from cable flexing. Instability can be avoided with small capacitor on input.

LOGARITHMIC AMPLIFIER

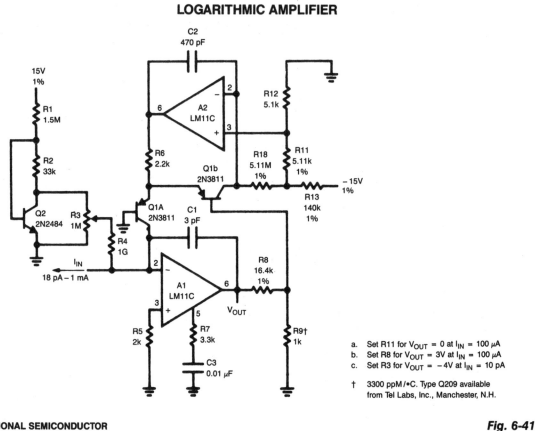

a. Set R11 for V_{OUT} = 0 at I_{IN} = 100 μA
b. Set R8 for V_{OUT} = 3V at I_{IN} = 100 μA
c. Set R3 for V_{OUT} = −4V at I_{IN} = 10 pA

† 3300 ppM/•C. Type Q209 available from Tel Labs, Inc., Manchester, N.H.

NATIONAL SEMICONDUCTOR

Fig. 6-41

Unusual frequency compensation gives this logarithmic converter a 100-μs time constant from 1 mA to 100 μA, increasing from 200 μs to 200 ms from 10 nA to 10 pA. Optional bias current compensation can give 10-pA resolution from −55 to 100°C. The scale factor is 1 V/decade, temperature compensated.

VOLTAGE-CONTROLLED VARIABLE-GAIN AMPLIFIER

The 2N5457 acts as a voltage variable resistor with an $R_{ds(on)}$ of 800 Ω max. Since the differential voltage on the LM101 is in the low-mV range, the 2N5457 JFET will have linear resistance over several decades of resistance providing an excellent electronic gain control.

NATIONAL SEMICONDUCTOR

Fig. 6-42

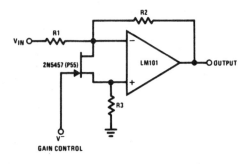

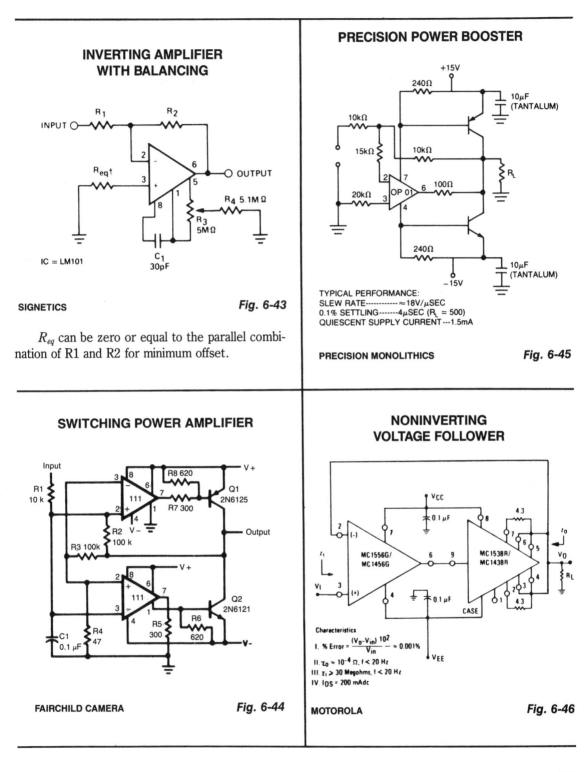

INVERTING AMPLIFIER WITH BALANCING

INPUT

R_1

R_2

$R_{eq}\dagger$

OUTPUT

R_4 5.1M Ω

R_3 5MΩ

C_1 30pF

IC = LM101

SIGNETICS

Fig. 6-43

R_{eq} can be zero or equal to the parallel combination of R1 and R2 for minimum offset.

PRECISION POWER BOOSTER

+15V

240Ω

10μF (TANTALUM)

10kΩ

15kΩ

10kΩ

R_L

20kΩ

OP 01

100Ω

240Ω

10μF (TANTALUM)

−15V

TYPICAL PERFORMANCE:
SLEW RATE------------ $\approx$18V/μSEC
0.1% SETTLING-------4μSEC (R_L = 500)
QUIESCENT SUPPLY CURRENT ---1.5mA

PRECISION MONOLITHICS

Fig. 6-45

SWITCHING POWER AMPLIFIER

Input

R1 10 k

R8 620

V +

Q1 2N6125

111

R7 300

V −

Output

R2 100 k

R3 100k

V +

111

Q2 2N6121

C1 0.1 μF

R4 47

R5 300

R6 620

V−

FAIRCHILD CAMERA

Fig. 6-44

NONINVERTING VOLTAGE FOLLOWER

V_{CC}

0.1 μF

4.3

z_i

MC1556G/ MC1456G

MC1538R/ MC1438R

V_O

V_I

R_L

0.1 μF

CASE

4.3

V_{EE}

Characteristics

I. % Error = $\frac{(V_0-V_{in})\ 10^2}{V_{in}}$ − $\approx$ 0.001%

II. $z_0 \approx 10^{-4}\ \Omega$, f < 20 Hz

III. $z_i >$ 30 Megohms, f < 20 Hz

IV. I_{OS} = 200 mAdc

MOTOROLA

Fig. 6-46

107

HIGH-IMPEDANCE DIFFERENTIAL AMPLIFIER

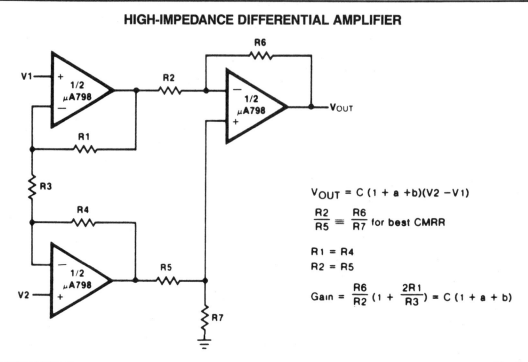

$$V_{OUT} = C\,(1 + a + b)(V2 - V1)$$

$$\frac{R2}{R5} \equiv \frac{R6}{R7} \text{ for best CMRR}$$

$$R1 = R4$$
$$R2 = R5$$

$$\text{Gain} = \frac{R6}{R2}\left(1 + \frac{2R1}{R3}\right) = C\,(1 + a + b)$$

FAIRCHILD CAMERA

Fig. 6-47

UNITY-GAIN FOLLOWER

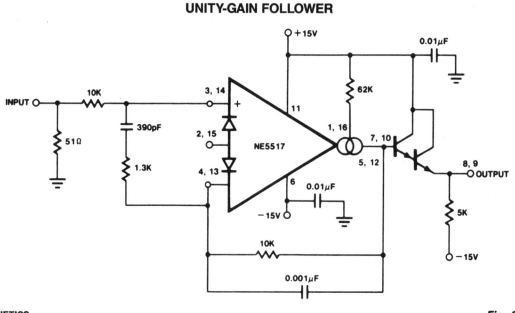

SIGNETICS

Fig. 6-48

VARIABLE-GAIN AND SIGN OP AMP

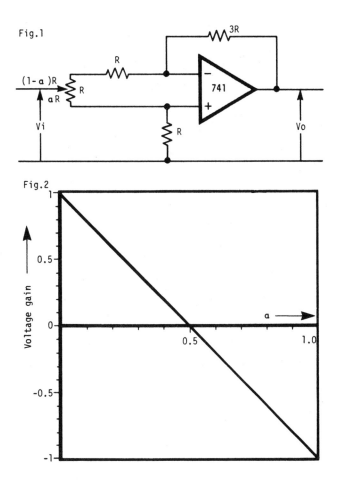

Fig.1

Fig.2

Fig. 6-49

The gain of the amplifier is smoothly controllable between the limits of +1 to −1. It is adjustable over this range using a single potentiometer. The voltage gain of the arrangement is given by:

$$\frac{V_o}{V_i} = \frac{2(1-2\alpha)}{(1+\alpha)(2-\alpha)}$$

Where α represents the fractional rotation of the potentiometer, R.

DISCRETE CURRENT BOOSTER

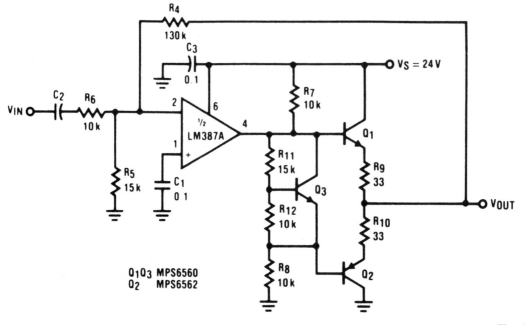

Fig. 6-50

PRECISION PROCESS-CONTROL INTERFACE

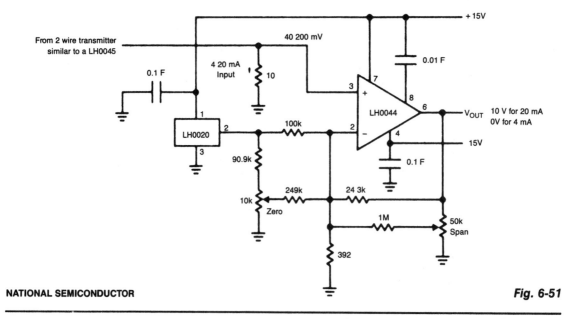

Fig. 6-51

INTRINSICALLY SAFE PROTECTED OP AMP

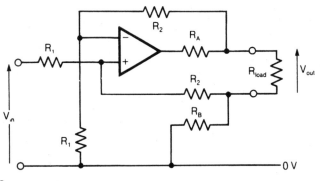

ELECTRONIC ENGINEERING

Fig. 6-52

In intrinsically safe applications, it is sometimes necessary to separate sections of circuitry by resistors which limit current under fault conditions. The circuit shown provides an accurate analogue output with effectively zero output impedance, despite having resistors in series with the output. The output voltage is given by:

$$V_{out} = \frac{V_{in}\,R_2}{R_1}$$

which is independent of R_A and R_B. The values of R_A and R_B should be chosen to achieve the desired current limiting, but note that a proportion of the voltage given at the op-amp output will be dropped across these resistors. This limits the output swing at the load to approximately:

$$\frac{V_S\,R_{load}}{R_A + R_B + R_{load}}$$

where: V_S = voltage swing at the op-amp output. Any type of op amp would be suitable.

±15-V CHOPPER AMPLIFIER

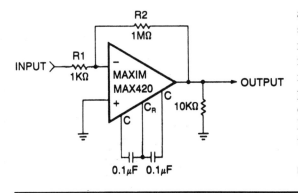

This simple circuit is a gain-of-1000 inverting amplifier. It will amplify submillivolt signals up to signal levels suitable for further processing. In almost all system applications, it is best to use as much gain as possible in the MAX420, thus minimizing the effects of later-stage offsets. For example, if circuitry following the MAX420 has an offset of 5 mV, the additional offset referred back to the MAX420 input (gain = 1000) will be 5 μV, doubling the system's offset error.

MAXIM

Fig. 6-53

INPUT/OUTPUT BUFFER AMPLIFIER FOR ANALOG MULTIPLEXERS

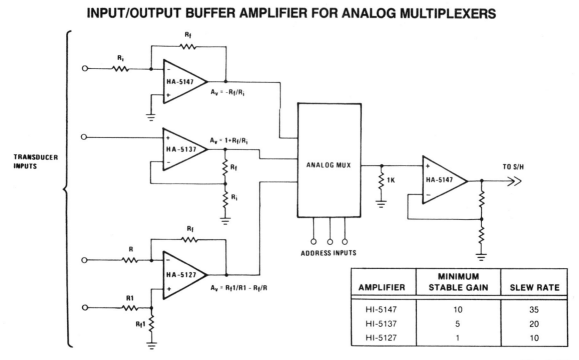

AMPLIFIER	MINIMUM STABLE GAIN	SLEW RATE
HI-5147	10	35
HI-5137	5	20
HI-5127	1	10

HARRIS

Fig. 6-54

The precision input characteristics of the HA-5147 help simplify system *error budgets*, while its speed and drive capabilities provide fast charging of the multiplexer's output capacitance. This speed eliminates an increased multiplexer acquisition time, which can be induced by more limited amplifiers. The HA-5147 accurately transfers information to the next stage while effectively reducing any loading effects on the multiplexer's output.

ABSOLUTE-VALUE NORTON AMPLIFIER

The noninverting amplifier has a gain of R2/R3 (1 in this case) and produces a voltage of V_{out} during a positive excursion of V_{in} with respect to ground. The inverting amplifier accommodates the negative excursions of V_{in}; its gain is given by $-$R6/R7, which equals -1 to maintain symmetry with the noninverting amplifier. R9 provides adjustment for the symmetry, supply variations, and offsets. Even though the circuit operates on a single supply, V_{in} can go negative to the same extent that it goes positive.

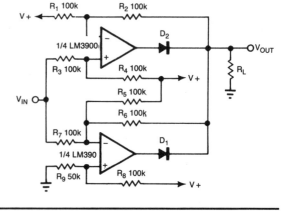

EDN

Fig. 6-55

HIGH-INPUT-IMPEDANCE DIFFERENTIAL AMPLIFIER

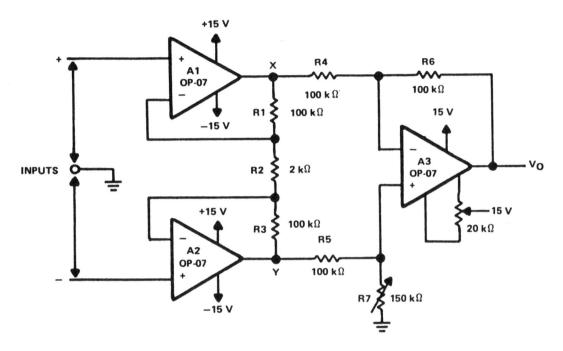

TEXAS INSTRUMENTS

Fig. 6-56

Operational amplifiers A1 and A2 are connected in a noninverting configuration and their outputs drive amplifier A3. Operational amplifier A3 could be called *subtractor circuit*, which converts the differential signal floating between points X and Y into a single-ended output voltage. Although not mandatory, amplifier A3 is usually operated at unity gain and R4, R5, R6, and R7 are all equal.

The common-mode-rejection of amplifier A3 is a function of how closely the ratio R4:R5 matches the ratio R6:R7. For example, when using resistors with 0.1% tolerance, common-mode rejection is greater than 60 dB. Additional improvement can be attained by using a potentiometer (slightly higher in value than R6) for R7. The potentiometer can be adjusted for the best common-mode rejection. Input amplifiers A1 and A2 will have some differential gain but the common-mode input voltages will experience only unity gain. These voltages will not appear as differential signals at the input of amplifier A3 because, when they appear at equal levels on both ends of resistor R2, they are effectively canceled.

This type of low-level differential amplifier finds widespread use in signal processing. It is also useful for dc and low-frequency signals that are commonly received from a transducer or thermocouple output, which are amplified and transmitted in a single-ended mode. The amplifier is powered by ± 15-V supplies. It is only necessary to null the input offset voltage of the output amplifier, A3.

HI-FI COMPANDOR

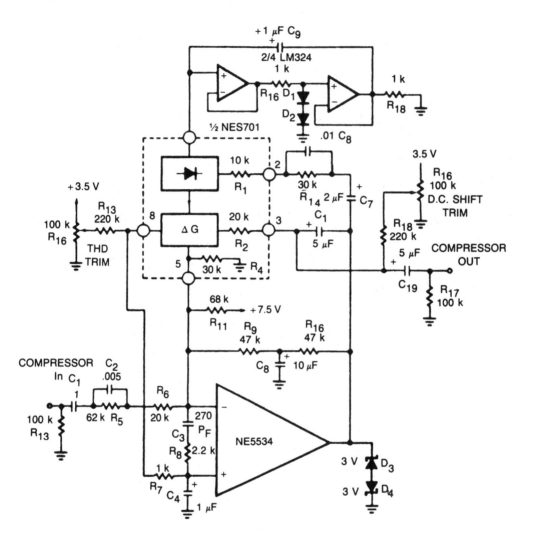

Fig. 6-57A

This circuit for a high-fidelity compressor uses an external op amp, and has high gain and wide bandwidth. An input compensation network is required for stability. The rectifier capacitor (C₉) is not grounded but it is tied to the output of an op-amp circuit. When a compressor is operating at high gain, (small input signal), and is suddenly hit with a signal, it will overload until it can reduce its gain. The time it takes for the compressor to recover from overload is determined by the rectifier capacitor C9. The expandor to

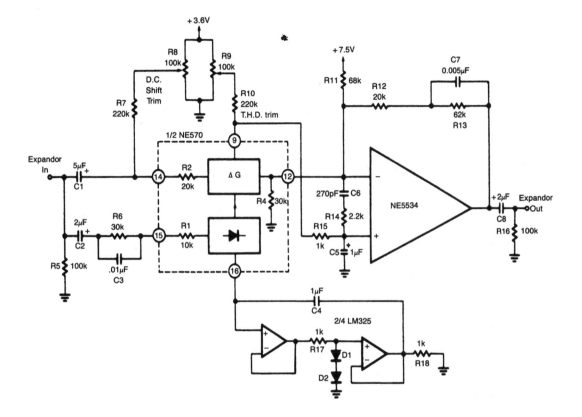

Fig. 6-57B

complement the compressor is shown in Fig. 6-57B. Here, an external op amp is used for high slew rate. Both the compressor and expandor have unity-gain levels of 0 dB. Trim networks are shown for distortion (THD) and dc shift. The distortion trim should be done first, with an input of 0 dB at 10 kHz. The dc shift should be adjusted for minimum envelope bounce with tone bursts.

ACTIVE-CLAMP LIMITING AMPLIFIER

The modified inverting amplifier uses an active clamp to limit the output swing with precision. Allowance must be made for the V_{BE} of the transistors. The swing is limited by the base-emitter breakdown of the transistors. A simple circuit uses two back-to-back zener diodes across the feedback resistor, but tends to give less precise limiting and cannot be easily controlled.

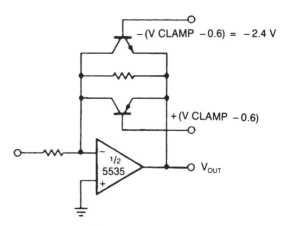

SIGNETICS

Fig. 6-58

WIDE-BAND AGC AMPLIFIER

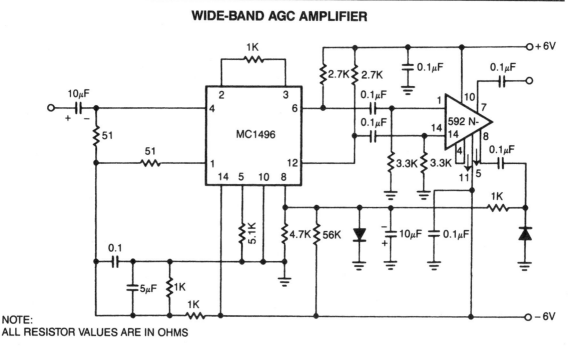

NOTE:
ALL RESISTOR VALUES ARE IN OHMS

SIGNETICS

Fig. 6-59

The NE592 is connected in conjunction with a MC1496 balanced modulator to form an excellent automatic gain control system. The signal is fed to the signal input of the MC1496 and rc-coupled to the NE592. Unbalancing the carrier input of the MC1496 causes the signal to pass through unattenuated. Rectifying and filtering one of the NE592 outputs produces a dc signal which is proportional to the ac signal amplitude. After filtering, this control signal is applied to the MC1496, causing its gain to change.

AUDIO AUTOMATIC GAIN CONTROL

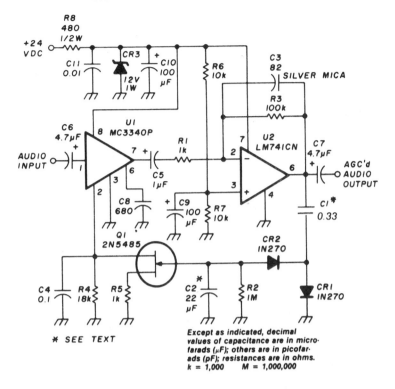

Fig. 6-60

An audio signal applied to U1 is passed through to the 741 operational amplifier, U2. After being amplified, the output signal of U2 is sampled and applied to a negative voltage-doubler/rectifier circuit that is composed of diodes CR1 and CR2, along with capacitor C1. The resulting negative voltage is used as a control voltage that is applied to the gate of the 2N5485 JFET Q1. Capacitor C2 and resistor R2 form a smoothing filter for the rectified audio control voltage.

The JFET is connected from pin 2 of the MC3340P to ground through a 1-kΩ resistor. As the voltage applied to the gate of the JFET becomes more negative in magnitude, the channel resistance of the JFET increases, causing the JFET to operate as a voltage-controlled resistor. The MC3340P audio attenuator is the heart of the AGC. It is capable of 13-dB gain (nearly −80 dB of attenuation), depending on the external resistance placed between pin 2 and ground. An increase of resistance decreases the gain that is achieved through the MC3340P. The circuit gain is not entirely a linear function of the external resistance, but it approximates such behavior over a good portion of the gain/attenuation range. An input signal applied to the AGC input will cause the gate voltage of the JFET to become proportionally negative. As a result, the JFET increases the resistance from pin 2 to ground of the MC3340P and causes a reduction in gain. In this way, the AGC output is held at a nearly constant level.

CHOPPER-STABILIZED AMPLIFIER

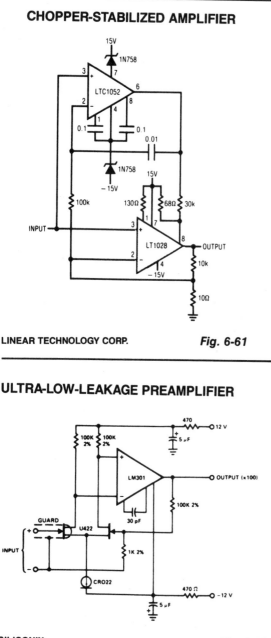

LINEAR TECHNOLOGY CORP.

Fig. 6-61

FET INPUT AMPLIFIER

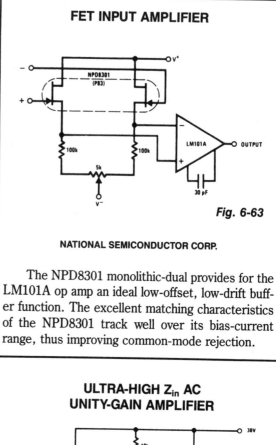

Fig. 6-63

NATIONAL SEMICONDUCTOR CORP.

The NPD8301 monolithic-dual provides for the LM101A op amp an ideal low-offset, low-drift buffer function. The excellent matching characteristics of the NPD8301 track well over its bias-current range, thus improving common-mode rejection.

ULTRA-LOW-LEAKAGE PREAMPLIFIER

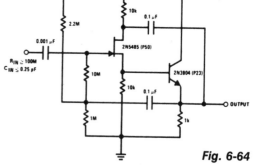

SILICONIX

Fig. 6-62

The circuit has an input leakage of only 2 pA typical at 75°C and would be usable with 1-MΩ input resistance.

ULTRA-HIGH Z_{in} AC
UNITY-GAIN AMPLIFIER

Fig. 6-64

NATIONAL SEMICONDUCTOR CORP.

Nothing is left to chance in reducing input capacitance. The 2N5485, which normally has low capacitance, is operated as a source follower with a bootstrapped gate-bias resistor and drain.

STEREO AMPLIFIER WITH GAIN CONTROL

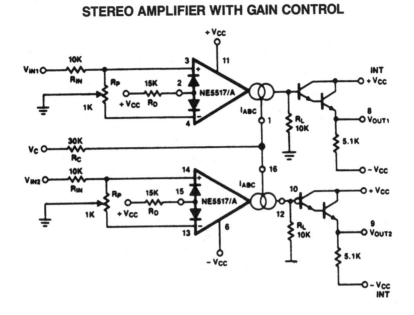

SIGNETICS

Fig. 6-65

Excellent tracking of 0.3 dB (typical) is easy to achieve. With the potentiometer, Rp, the offset can be adjusted. For ac-coupled amplifiers, the potentiometer can be replaced with two 5.1-kΩ resistors.

NONINVERTING AMPLIFIER

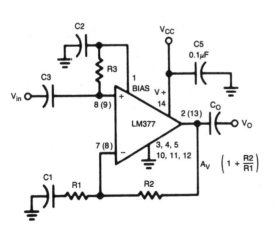

$$A_V \quad \left(1 + \frac{R2}{R1}\right)$$

Fig. 6-66

NATIONAL SEMICONDUCTOR

7

Programmable Amplifiers

The sources of the following circuits are contained in the Sources section, which begins on page 182. The figure number contained with each circuit correlates to the source entry in the Sources section.

INVERTING PROGRAMMABLE-GAIN AMPLIFIER

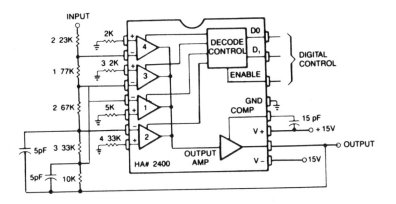

HARRIS

Fig. 7-1

This circuit can be programmed for a gain of 0, −1, −2, −4, or −8. This could also be accomplished with one input resistor and one feedback resistor per channel in the conventional manner, but this would require eight resistors, rather than five.

NONINVERTING PROGRAMMABLE-GAIN AMPLIFIER

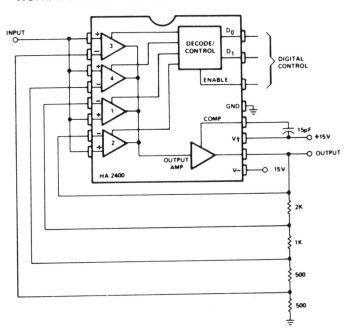

HARRIS

Fig. 7-2

This is a noninverting amplifier configuration with feedback resistors chosen to produce a gain of 0, 1, 2, 4, or 8, depending on the digital control inputs. Comparators at the output could be used for automatic gain selection for auto-ranging meters, etc.

WIDE-RANGE DIGITALLY CONTROLLED VARIABLE-GAIN AMPLIFIER

The circuit uses the LTC1043 in a variable gain amplifier which features continuously variable gain, gain stability of 20 ppm/°C, and single-ended or differential inputs. The circuit uses two separate LTC1043s. LTC1043B is continuously clocked by a 1-kHz source, which could also be processor supplied. Both LTC1043s function as the sampled data equivalent of a resistor within the bandwidth set by A1's 0.01-μF value and the switched-capacitor equivalent feedback resistor. The time-averaged current delivered to the summing point by LTC1043A is a function of the 0.01-μF capacitor's input-derived voltage and the commutation frequency at pin 16. Low-commutation frequencies result in small time-averaged current values, and require a large input resistor. Higher frequencies require an equivalent small input resistor.

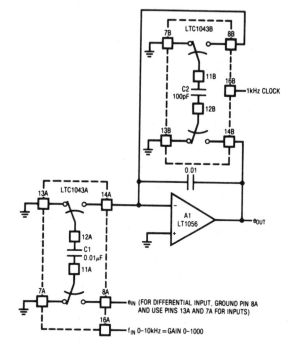

LINEAR TECHNOLOGY CORP.

Fig. 7-3

DIGITALLY PROGRAMMABLE PRECISION AMPLIFIER

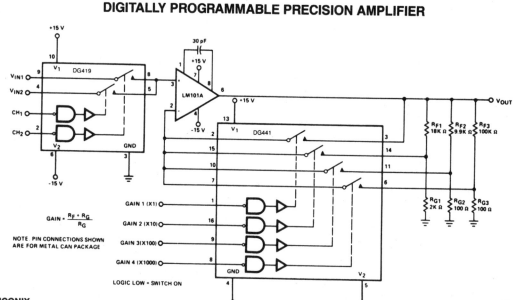

SILICONIX

Fig. 7-4

The DG419 *looks* into the high input impedance of the op amp, so the effects of $R_{DS(on)}$ are negligible. The DG441 is also connected in series with R_{IN} and is not included in the feedback dividers, thus contributing negligible error to the overall gain. Because the DG419 and DG441 can handle ± 15 V, the unity gain follower connection, X1, is capable of the full op-amp output range of ± 12 V.

PROGRAMMABLE-GAIN DIFFERENTIAL-INPUT AMPLIFIER

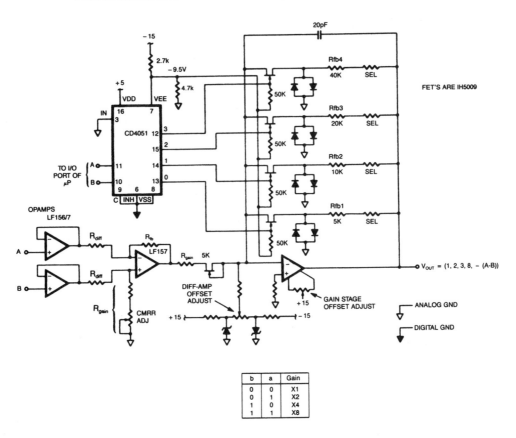

b	a	Gain
0	0	X1
0	1	X2
1	0	X4
1	1	X8

INTERSIL

Fig. 7-5

This programmable gain circuit employs a CD4051 CMOS Analog Multiplexer as a two-to-four line decoder, with appropriate FET drive for switching between feedback resistors to program the gain to any one of four values.

PROGRAMMABLE-GAIN NONINVERTING
AMPLIFIER WITH SELECTABLE INPUTS

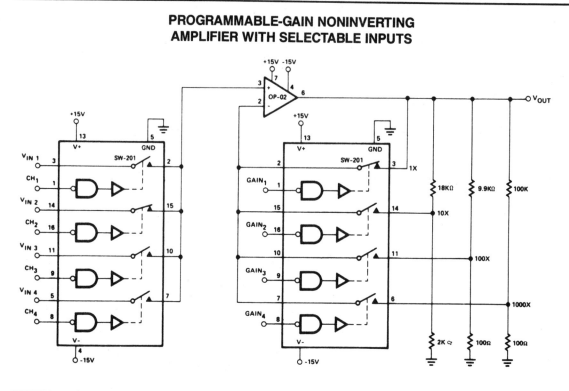

PRECISION MONOLITHICS

Fig. 7-6

PROGRAMMABLE AMPLIFIER

Often a circuit will be called upon to perform several functions. In these situations, the variable gain configuration of this circuit could be quite useful. This programmable gain stage depends on CMOS analog switches to alter the amount of feedback, and thereby, the gain of the stage. Placement of the switching elements inside the relatively low-current area of the feedback loop, minimizes the effects of bias currents and switch resistance on the calculated gain of the stage. Voltage spikes can occur during the switching process, resulting in temporarily reduced gain because of the make-before-break operation of the switches. This gain loss can be minimized by providing a separate voltage divider network for each level of gain.

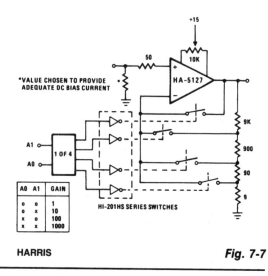

HARRIS

Fig. 7-7

PRECISION-WEIGHTED RESISTOR PROGRAMMABLE-GAIN AMPLIFIER

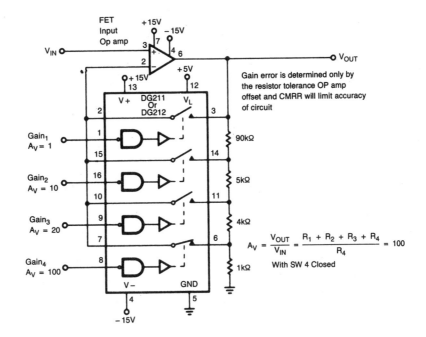

Fig. 7-8

8

RF Amplifiers

The sources of the following circuits are contained in the Sources section, which begins on page 182. The figure number contained with each circuit correlates to the source entry in the Sources section.

5-W 150-MHz AMPLIFIER

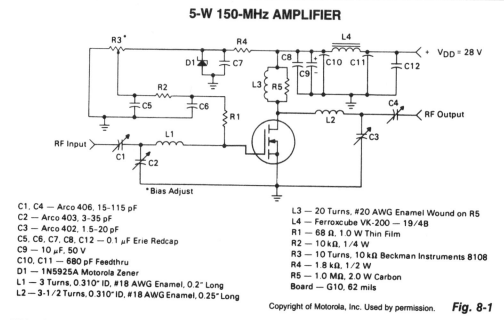

C1, C4 — Arco 406, 15-115 pF
C2 — Arco 403, 3-35 pF
C3 — Arco 402, 1.5-20 pF
C5, C6, C7, C8, C12 — 0.1 μF Erie Redcap
C9 — 10 μF, 50 V
C10, C11 — 680 pF Feedthru
D1 — 1N5925A Motorola Zener
L1 — 3 Turns, 0.310″ ID, #18 AWG Enamel, 0.2″ Long
L2 — 3-1/2 Turns, 0.310″ ID, #18 AWG Enamel, 0.25″ Long

L3 — 20 Turns, #20 AWG Enamel Wound on R5
L4 — Ferroxcube VK-200 — 19/4B
R1 — 68 Ω, 1.0 W Thin Film
R2 — 10 kΩ, 1/4 W
R3 — 10 Turns, 10 kΩ Beckman Instruments 8108
R4 — 1.8 kΩ, 1/2 W
R5 — 1.0 MΩ, 2.0 W Carbon
Board — G10, 62 mils

Fig. 8-1

This circuit utilizes the MRF123 TMOS power FET. The MRF134 is a very high gain FET that is potentially unstable at both VHF and UHF frequencies. Note that a 68-Ω input loading resistor has been utilized to enhance stability. This amplifier has a gain of 14 dB and a drain efficiency of 55%.

UHF-TV PREAMPLIFIER

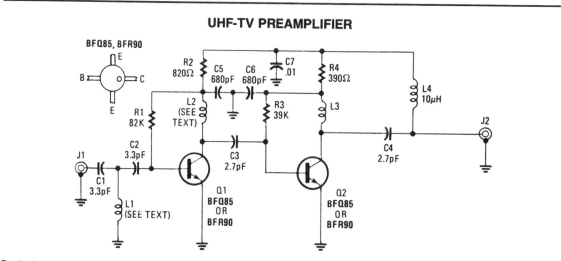

Fig. 8-2

An inexpensive, antenna-mounted, UHF-TV preamplifier can add more than 25 dB of gain. The first stage of the preamp is biased for optimum noise, the second stage for optimum gain. L1, L2 strip line ≈ λ/8 part of PC board.

128

1-W 2.3-GHz AMPLIFIER

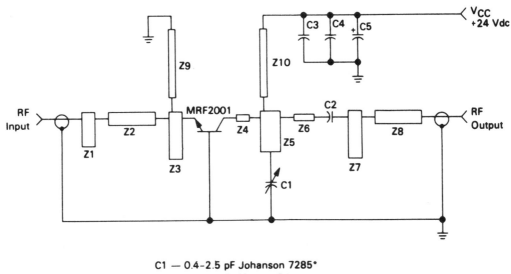

C1 — 0.4-2.5 pF Johanson 7285*
C2, C3 — 68 pF, 50 mil ATC**
C4 — 0.1 μF, 50 V
C5 — 4.7 μF, 50 V Tantalum

Z1-Z10 — Microstrip; see Photomaster, Figure 3

Board Material — 0.0625" 3M Glass Teflon,***
ϵ_r = 2.5 ± 0.05
*Johanson Manufacturing Corp., 400 Rockaway Valley Road, Boonton, NJ 07005
**American Technical Ceramics, One Norden Lane, Huntington Station, NY 11746
***Registered Trademark of Du Pont

MOTOROLA

Fig. 8-3

Simplicity and repeatability are featured in this 1-W S-band amplifier design. The design uses an MRF2001 transistor as a common-base Class-C amplifier. The amplifier delivers 1-W output with 8-dB minimum gain at 24 V, and is tunable from 2.25 to 2.35 GHz. Applications include microwave communications equipment and other systems that require medium-power narrow-band amplification. The amplifier circuitry consists almost entirely of distributed microstrip elements. A total of six additional components, including the MRF2001, are required to build a working amplifier. The input and output impedances of the transistor are matched to 50 ohms by double section low-pass networks. The networks are designed to provide about 3% 1-dB power bandwidth, while maintaining a collector efficiency of approximately 30%. The only tuning adjustment is in C1 in the output network. Ceramic-chip capacitors, C2 and C3, are used for dc blocking and power-supply decoupling. Additional low-frequency decoupling is provided by capacitors C4 and C5.

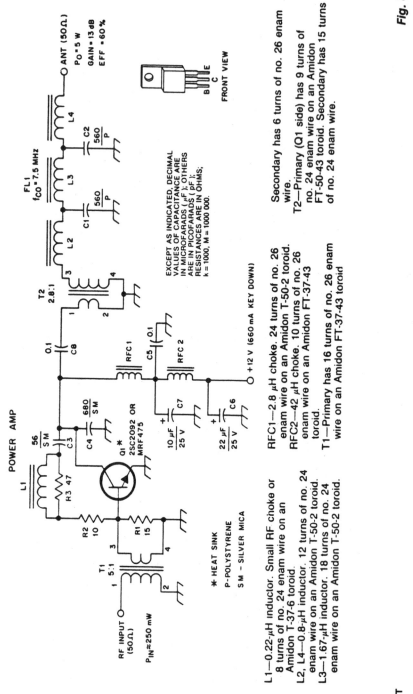

5-W RF POWER AMPLIFIER

Fig. 8-4

L1—0.22-μH inductor. Small RF choke or 8 turns of no. 24 enam wire on an Amidon T-37-6 toroid.
L2, L4—0.8-μH inductor. 12 turns of no. 24 enam wire on an Amidon T-50-2 toroid.
L3—1.67-μH inductor. 18 turns of no. 24 enam wire on an Amidon T-50-2 toroid.

RFC1—2.8 μH choke. 24 turns of no. 26 enam wire on an Amidon T-50-2 toroid.
RFC2—42 μH choke. 10 turns of no. 26 enam wire on an Amidon FT-37-43 toroid.
T1—Primary has 16 turns of no. 26 enam wire on an Amidon FT-37-43 toroid

Secondary has 6 turns of no. 26 enam wire.
T2—Primary (Q1 side) has 9 turns of no. 24 enam wire on an Amidon FT-50-43 toroid. Secondary has 15 turns of no. 24 enam wire.

Numbered components are designated for PC-board layout purposes. C5 and C8 are disc ceramic. C6 and C7 are tantalum or electrolytic. R1, R2, and R3 are 1/2-W carbon-composition resistors. Silver-mica capacitors can be substituted for polystyrene (P) types. Impedance transformation ratios are shown above T1 and T2.

QST

LOW-NOISE BROADBAND AMPLIFIER

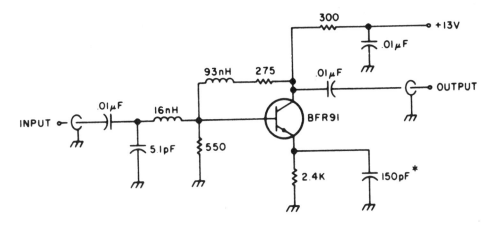

ELECTRONICS TODAY INTERNATIONAL

Fig. 8-5

The amplifier provides 10 dB of gain from 10-600 MHz and has a 1.5-to-1 match at 50 ohms. The BFR91 has a 1.5-dB noise figures at 500 MHz. The circuit requires 13.5 Vdc at about 13 mA. Keep the leads on the 150-pF emitter bypass capacitor as short as possible. The 16-nH coil is 2.5 turns of #26 enamel wire on the shank of a #40 drill. The 93-nH inductor is 10 turns of the same material.

2-METER 10-W POWER AMPLIFIER

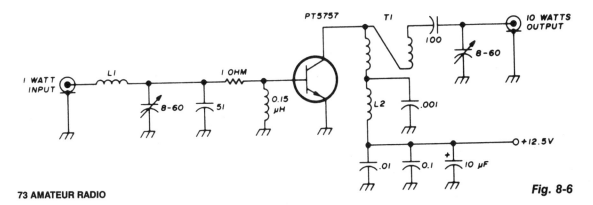

73 AMATEUR RADIO

Fig. 8-6

This 10-W 144-MHz power amplifier uses a TRW PT5757 transistor. L1 is 4 turns of #20 enameled, 3/32″ ID; L2 is 10 turns of #20 enameled, 3/32″ ID. Transformer T1 is a 4:1 transmission-line transformer, made from a 3″ length of twisted pair of #20 enameled wire.

10-W 225- to 400-MHz AMPLIFIER

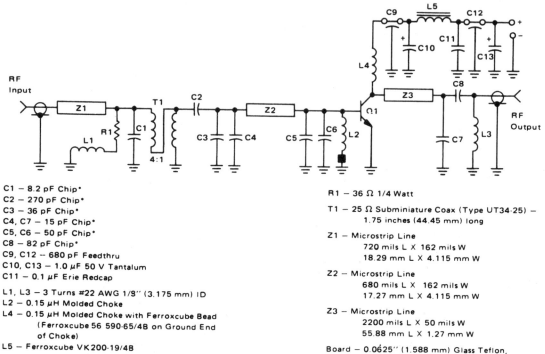

C1 — 8.2 pF Chip*
C2 — 270 pF Chip*
C3 — 36 pF Chip*
C4, C7 — 15 pF Chip*
C5, C6 — 50 pF Chip*
C8 — 82 pF Chip*
C9, C12 — 680 pF Feedthru
C10, C13 — 1.0 µF 50 V Tantalum
C11 — 0.1 µF Erie Redcap

L1, L3 — 3 Turns #22 AWG 1/8" (3.175 mm) ID
L2 — 0.15 µH Molded Choke
L4 — 0.15 µH Molded Choke with Ferroxcube Bead
    (Ferroxcube 56 590-65/4B on Ground End
    of Choke)
L5 — Ferroxcube VK 200-19/4B

*100 mil A.C.I. Chip Capacitors

R1 — 36 Ω 1/4 Watt

T1 — 25 Ω Subminiature Coax (Type UT34-25) —
    1.75 inches (44.45 mm) long

Z1 — Microstrip Line
    720 mils L X 162 mils W
    18.29 mm L X 4.115 mm W

Z2 — Microstrip Line
    680 mils L X 162 mils W
    17.27 mm L X 4.115 mm W

Z3 — Microstrip Line
    2200 mils L X 50 mils W
    55.88 mm L X 1.27 mm W

Board — 0.0625" (1.588 mm) Glass Teflon,
    ϵ_r = 2.56

Q1 — MRF331

SCHEMATIC REPRESENTATION

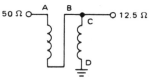

ASSEMBLY AND PICTORIAL

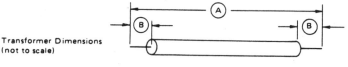

Transformer Dimensions
(not to scale)

Ⓐ — 1.75 inches (4.445 cm)

Ⓑ — 0.1875 inch (0.476 cm)

Transformer Connections

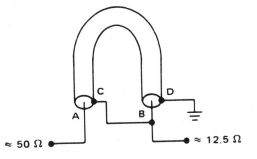

$\approx 50\ \Omega$

$\approx 12.5\ \Omega$

MOTOROLA

Fig. 8-7

This broadband amplifier covers the 225- to 400-MHz military-communications band producing 10-W RF output power and operating from a 28-V supply. The amplifier can be used as a driver for higher power devices, such as 2N6439 and MRF327. The circuit is designed to be driven by a 50-Ω source and operate into a nominal 50-Ω load. The input matching network consists of a section composed of C3, C4, Z2, C5, and C6. C2 is a dc-blocking capacitor and T1 is a 4:1 impedance-ratio coaxial transformer. Z1 is a 50-Ω transmission line. A compensation network that consists of R1, C1, and L1 is used to improve the input VSWR and flatten the gain response of the amplifier. L2 and a small ferrite bead make up the base bias choke. The output network is made up of a microstrip L-section that consists of Z3 and C7, and a high-pass section that consists of C8 and L3. C8 also serves as a dc-blocking capacitor. Collector decoupling is accomplished through the use of L4, L5, C9, C10, C11, C12, and C13.

60-W 225- to 400-MHz AMPLIFIER

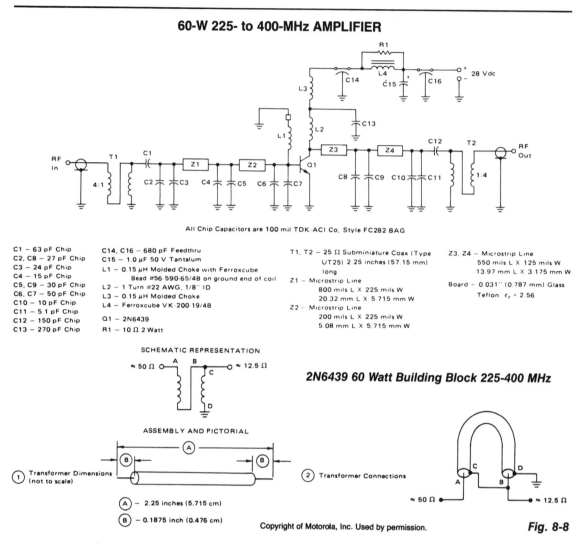

All Chip Capacitors are 100 mil TDK-ACI Co. Style FC282 BAG

C1 – 63 pF Chip
C2, C8 – 27 pF Chip
C3 – 24 pF Chip
C4 – 15 pF Chip
C5, C9 – 30 pF Chip
C6, C7 – 50 pF Chip
C10 – 10 pF Chip
C11 – 5.1 pF Chip
C12 – 150 pF Chip
C13 – 270 pF Chip

C14, C16 – 680 pF Feedthru
C15 – 1.0 μF 50 V Tantalum
L1 – 0.15 μH Molded Choke with Ferroxcube
  Bead #56-590-65/4B on ground end of coil
L2 – 1 Turn #22 AWG, 1/8" ID
L3 – 0.15 μH Molded Choke
L4 – Ferroxcube VK-200-19/4B
Q1 – 2N6439
R1 – 10 Ω 2 Watt

T1, T2 – 25 Ω Subminiature Coax (Type
  UT25) 2.25 inches (57.15 mm)
  long
Z1 – Microstrip Line
  800 mils L X 225 mils W
  20.32 mm L X 5.715 mm W
Z2 – Microstrip Line
  200 mils L X 225 mils W
  5.08 mm L X 5.715 mm W

Z3, Z4 – Microstrip Line
  550 mils L X 125 mils W
  13.97 mm L X 3.175 mm W

Board – 0.031" (0.787 mm) Glass
  Teflon ϵ_r = 2.56

2N6439 60 Watt Building Block 225-400 MHz

Copyright of Motorola, Inc. Used by permission.

Fig. 8-8

Construction Details of the 4:1 Unbalanced to Unbalanced Transformers

This 60-W, 28-V broadband amplifier covers the 225 – 400 MHz military communications band. The amplifier may be used singly as a 60-W output stage in a 225 – 400 MHz transmitter, or by using two of these amplifiers combined with quadrature couplers, a 100-W output amplifier stage can be constructed. The circuit is designed to be driven from a 50-Ω source and work into a nominal 50-Ω load. The input network consists of two microstrip L-sections composed of Z1, Z2, and C2 through C6. C1 serves as a dc-blocking capacitor. A 4:1 impedance ratio coaxial transformer T1 completes the input matching network. L1 and a ferrite bead serve as a base decoupling choke. The output circuit consists of shunt inductor L2 at the collector, followed by two microstrip L-sections composed of Z3, Z4, and C8 through C11. C12 serves as a dc blocking capacitor, and is followed by another 4:1 impedance ratio coaxial transformer. Collector decoupling is accomplished through the use of L3, L4, C14, C15, C16, and R1.

SINGLE-DEVICE 80-W 50-Ω VHF AMPLIFIER

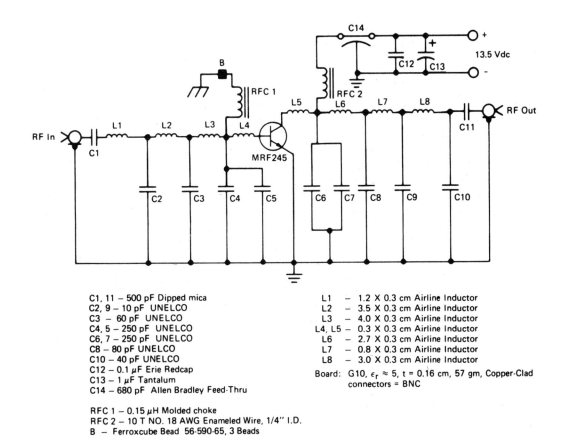

C1, 11 – 500 pF Dipped mica
C2, 9 – 10 pF UNELCO
C3 – 60 pF UNELCO
C4, 5 – 250 pF UNELCO
C6, 7 – 250 pF UNELCO
C8 – 80 pF UNELCO
C10 – 40 pF UNELCO
C12 – 0.1 μF Erie Redcap
C13 – 1 μF Tantalum
C14 – 680 pF Allen Bradley Feed-Thru

RFC 1 – 0.15 μH Molded choke
RFC 2 – 10 T NO. 18 AWG Enameled Wire, 1/4″ I.D.
B – Ferroxcube Bead 56-590-65, 3 Beads

L1   – 1.2 X 0.3 cm Airline Inductor
L2   – 3.5 X 0.3 cm Airline Inductor
L3   – 4.0 X 0.3 cm Airline Inductor
L4, L5 – 0.3 X 0.3 cm Airline Inductor
L6   – 2.7 X 0.3 cm Airline Inductor
L7   – 0.8 X 0.3 cm Airline Inductor
L8   – 3.0 X 0.3 cm Airline Inductor

Board: G10, $\epsilon_r \approx 5$, t = 0.16 cm, 57 gm, Copper-Clad
connectors = BNC

MOTOROLA

Fig. 8-9

The amplifier uses a single MRF245 and provides 80 W with 9.4-dB gain across 143- to 156-MHz.

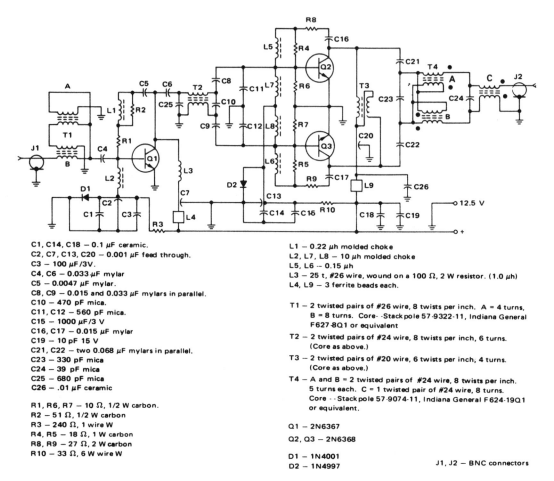

C1, C14, C18 — 0.1 µF ceramic.
C2, C7, C13, C20 — 0.001 µF feed through.
C3 — 100 µF/3V.
C4, C6 — 0.033 µF mylar
C5 — 0.0047 µF mylar.
C8, C9 — 0.015 and 0.033 µF mylars in parallel.
C10 — 470 pF mica.
C11, C12 — 560 pF mica.
C15 — 1000 µF/3 V
C16, C17 — 0.015 µF mylar
C19 — 10 pF 15 V
C21, C22 — two 0.068 µF mylars in parallel.
C23 — 330 pF mica
C24 — 39 pF mica
C25 — 680 pF mica
C26 — .01 µF ceramic

R1, R6, R7 — 10 Ω, 1/2 W carbon.
R2 — 51 Ω, 1/2 W carbon
R3 — 240 Ω, 1 wire W
R4, R5 — 18 Ω, 1 W carbon
R8, R9 — 27 Ω, 2 W carbon
R10 — 33 Ω, 6 W wire W

L1 — 0.22 µh molded choke
L2, L7, L8 — 10 µh molded choke
L5, L6 — 0.15 µh
L3 — 25 t, #26 wire, wound on a 100 Ω, 2 W resistor. (1.0 µh)
L4, L9 — 3 ferrite beads each.

T1 — 2 twisted pairs of #26 wire, 8 twists per inch. A = 4 turns,
B = 8 turns. Core - Stackpole 57-9322-11, Indiana General
F627-8Q1 or equivalent

T2 — 2 twisted pairs of #24 wire, 8 twists per inch, 6 turns.
(Core as above.)

T3 — 2 twisted pairs of #20 wire, 6 twists per inch, 4 turns.
(Core as above.)

T4 — A and B = 2 twisted pairs of #24 wire, 8 twists per inch.
5 turns each. C = 1 twisted pair of #24 wire, 8 turns.
Core - Stackpole 57-9074-11, Indiana General F624-19Q1
or equivalent.

Q1 — 2N6367

Q2, Q3 — 2N6368

D1 — 1N4001
D2 — 1N4997

J1, J2 — BNC connectors

Fig. 8-10

This amplifier utilizes a 2N6367 and a pair of 2N6368 transistors. The 2N6367 transistor is employed as a driver and is specified for up to 9-W (PEP) output. In the amplifier design the driver must supply 5 W (PEP) at 30 MHz with a resulting IMD performance of about −37 to −38 dB. At lower operating frequencies, drive requirements drop to the 2- to 3-W (PEP) range and IMD performance improves to better than 40 dB. Two 2N6368 transistors are employed in the final stage of the transmitter design in a push-pull configuration. These devices are rated at 40 W (PEP) and −30 dB maximum IMD, although −35-dB performance is more typical for narrowband operation. Without frequency compensation, the completed amplifier can deliver 90 W (PEP) in the 25- to 30-MHz band with IMD performance down −30 dB. If only the power amplifier stage is frequency compensated, 95 W (PEP) can be obtained at 6 to 10 MHz.

100-W PEP 420- to 450-MHz PUSH-PULL LINEAR AMPLIFIER

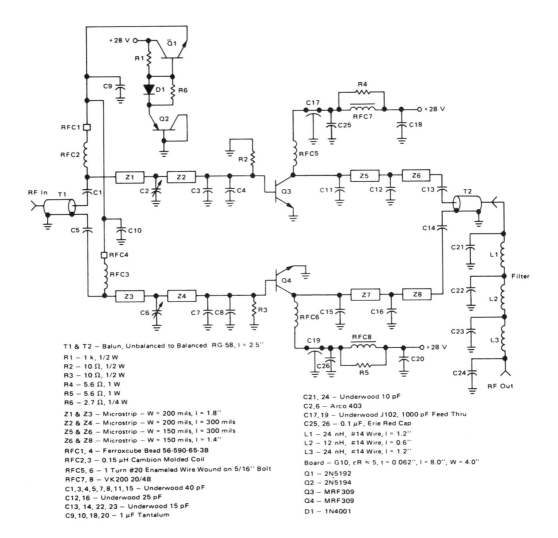

T1 & T2 – Balun, Unbalanced to Balanced RG-58, l = 2.5"

R1 – 1 k, 1/2 W
R2 – 10 Ω, 1/2 W
R3 – 10 Ω, 1/2 W
R4 – 5.6 Ω, 1 W
R5 – 5.6 Ω, 1 W
R6 – 2.7 Ω, 1/4 W

Z1 & Z3 – Microstrip – W = 200 mils, l = 1.8"
Z2 & Z4 – Microstrip – W = 200 mils, l = 300 mils
Z5 & Z6 – Microstrip – W = 150 mils, l = 300 mils
Z6 & Z8 – Microstrip – W = 150 mils, l = 1.4"
RFC1, 4 – Ferroxcube Bead 56-590-65-3B
RFC2,3 – 0.15 μH Cambion Molded Coil
RFC5, 6 – 1 Turn #20 Enameled Wire Wound on 5/16" Bolt
RFC7, 8 – VK200 20/4B
C1,3,4,5,7,8,11,15 – Underwood 40 pF
C12,16 – Underwood 25 pF
C13, 14, 22, 23 – Underwood 15 pF
C9,10,18,20 – 1 μF Tantalum

C21, 24 – Underwood 10 pF
C2,6 – Arco 403
C17, 19 – Underwood J102, 1000 pF Feed Thru
C25, 26 – 0.1 μF, Erie Red Cap
L1 – 24 nH, #14 Wire, l = 1.2"
L2 – 12 nH, #14 Wire, l = 0.6"
L3 – 24 nH, #14 Wire, l = 1.2"
Board – G10, εR ≈ 5, t = 0.062", l = 8.0", W = 4.0"
Q1 – 2N5192
Q2 – 2N5194
Q3 – MRF309
Q4 – MRF309
D1 – 1N4001

MOTOROLA

Fig. 8-11

This 100-W linear amplifier can be constructed using two MRF309 transistors in push-pull, requiring only 16-W drive from 420 to 450 MHz. Operating from a 28-V supply, eight dB of power gain is achieved, along with excellent practical performance which features: maximum input SWR of 2:1, harmonic suppression more than −63 dB below 100-W output, efficiency greater than 40%, and circuit stability with a 3:1 collector mismatch at all phase angles.

125-W 150-MHz AMPLIFIER

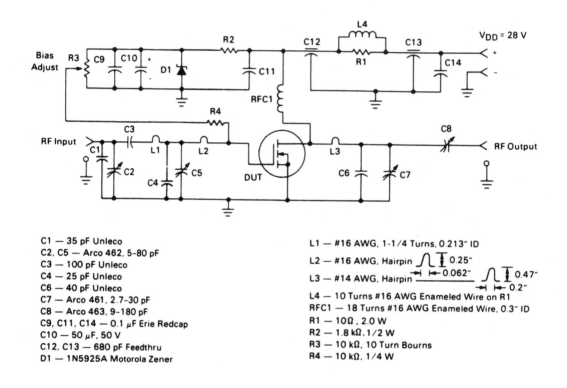

C1 — 35 pF Unleco
C2, C5 — Arco 462, 5–80 pF
C3 — 100 pF Unleco
C4 — 25 pF Unleco
C6 — 40 pF Unleco
C7 — Arco 461, 2.7–30 pF
C8 — Arco 463, 9–180 pF
C9, C11, C14 — 0.1 μF Erie Redcap
C10 — 50 μF, 50 V
C12, C13 — 680 pF Feedthru
D1 — 1N5925A Motorola Zener

L1 — #16 AWG, 1-1/4 Turns, 0.213″ ID
L2 — #16 AWG, Hairpin 0.25″
L3 — #14 AWG, Hairpin 0.062″ 0.47″ 0.2″
L4 — 10 Turns #16 AWG Enameled Wire on R1
RFC1 — 18 Turns #16 AWG Enameled Wire, 0.3″ ID
R1 — 10Ω, 2.0 W
R2 — 1.8 kΩ, 1/2 W
R3 — 10 kΩ, 10 Turn Bourns
R4 — 10 kΩ, 1/4 W

MOTOROLA

Fig. 8-12

This amplifier operates from a 28-Vdc supply. It has a typical gain of 12 dB, and it can survive operation into a 30:1 VSWR load at any phase angle with no damage. The amplifier has an AGC range in excess of 20 dB. So, with input power held constant at the level that provides 125-W output, the output power can be reduced to less than 1.0 W continuously by driving the dc gate voltage negative from its I_{DQ} value.

2- TO 30-MHz 140-W (PEP) AMATEUR-RADIO LINEAR AMPLIFIER

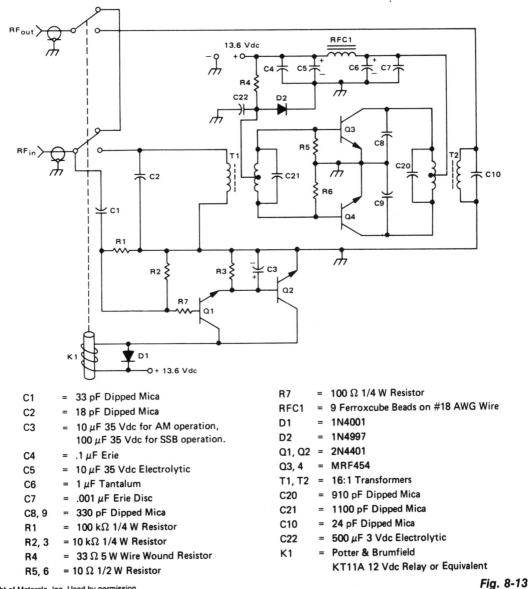

C1	=	33 pF Dipped Mica	R7	=	100 Ω 1/4 W Resistor
C2	=	18 pF Dipped Mica	RFC1	=	9 Ferroxcube Beads on #18 AWG Wire
C3	=	10 μF 35 Vdc for AM operation,	D1	=	1N4001
		100 μF 35 Vdc for SSB operation.	D2	=	1N4997
C4	=	.1 μF Erie	Q1, Q2	=	2N4401
C5	=	10 μF 35 Vdc Electrolytic	Q3, 4	=	MRF454
C6	=	1 μF Tantalum	T1, T2	=	16:1 Transformers
C7	=	.001 μF Erie Disc	C20	=	910 pF Dipped Mica
C8, 9	=	330 pF Dipped Mica	C21	=	1100 pF Dipped Mica
R1	=	100 kΩ 1/4 W Resistor	C10	=	24 pF Dipped Mica
R2, 3	=	10 kΩ 1/4 W Resistor	C22	=	500 μF 3 Vdc Electrolytic
R4	=	33 Ω 5 W Wire Wound Resistor	K1	=	Potter & Brumfield
R5, 6	=	10 Ω 1/2 W Resistor			KT11A 12 Vdc Relay or Equivalent

Fig. 8-13

The amplifier operates across the 2- to 30-MHz band with relatively flat gain response and reaches gain saturation at approximately 210 W of output power. Both input and output transformers are 4:1 turns ratio (16:1 impedance ratio) to achieve low input SWR across the specified band and a high saturation capability. When using this design, it is important to interconnect the ground plane on the bottom of the board to the top, especially at the emitters of the MRF454s.

160-W (PEP) BROADBAND LINEAR AMPLIFIER

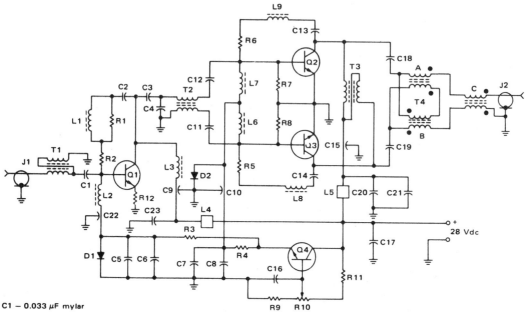

C1 — 0.033 μF mylar

C2, C3 — 0.01 μF mylar

C4 — 620 pF dipped mica

C5, C7, C16 — 0.1 μF ceramic

C6 — 100 μF/15 V electrolytic

C8 — 500 μF/6 V electrolytic

C9, C10, C15, C22 — 1000 pF feed through

C11, C12 — 0.01 μF

C13, C14 — 0.015 μF mylar

C17 — 10 μF/35 V electrolytic

C18, C19, C21 — Two 0.068 μF mylars in parallel

C20 — 0.1 μF disc ceramic

C23 — 0.1 μF disc ceramic

R1 — 220 Ω, 1/4 W carbon

R2 — 47 Ω, 1/2 W carbon

R3 — 820 Ω, 1 W wire W

R4 — 35 Ω, 5 W wire W

R5, R6 — Two 150 Ω, 1/2 W carbon in parallel

R7, R8 — 10 Ω, 1/2 W carbon

R9, R11 — 1 k, 1/2 W carbon

R10 — 1 k, 1/2 W potentiometer

R12 — 0.85 Ω (6 5.1 Ω or 4 3.3 Ω 1/4 W resistors in parallel, divided equally between both emitter leads)

T1 — 4:1 Transformer, 6 turns, 2 twisted pairs of #26 AWG enameled wire (8 twists per inch)

T2 — 1:1 Balun, 6 turns, 2 twisted pairs of #24 AWG enameled wire (6 twists per inch)

T3 — Collector choke, 4 turns, 2 twisted pairs of #22 AWG enameled wire (6 twists per inch)

T4 — 1:4 Transformer Balun, A&B — 5 turns, 2 twisted pairs of #24, C — 8 turns, 1 twisted pair of #24 AWG enameled wire (All windings 6 twists per inch). (T4 — Indiana General F624-19Q1, — All others are Indiana General F627-8Q1 ferrite toroids or equivalent.)

PARTS LIST

L1 — .33 μH, molded choke	Q1 -- 2N6370
L2, L6, L7 — 10 μH, molded choke	Q2, Q3 — 2N5942
L3 — 1.8 μH (Ohmite 2-144)	Q4 — 2N5190
L4, L5 — 3 ferrite beads each	D1 — 1N4001
L8, L9 — .22 μH, molded choke	D2 — 1N4997
	J1, J2 — BNC connectors

Fig. 8-14

600-W RF POWER AMPLIFIER

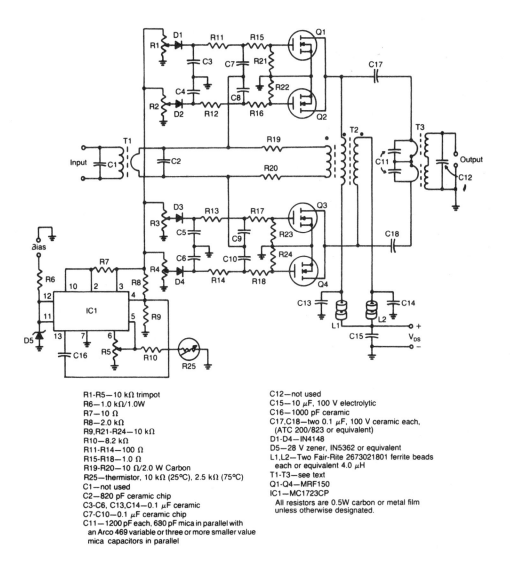

R1-R5—10 kΩ trimpot
R6—1.0 kΩ/1.0W
R7—10 Ω
R8—2.0 kΩ
R9,R21-R24—10 kΩ
R10—8.2 kΩ
R11-R14—100 Ω
R15-R18—1.0 Ω
R19-R20—10 Ω/2.0 W Carbon
R25—thermistor, 10 kΩ (25°C), 2.5 kΩ (75°C)
C1—not used
C2—820 pF ceramic chip
C3-C6, C13,C14—0.1 μF ceramic
C7-C10—0.1 μF ceramic chip
C11—1200 pF each, 680 pF mica in parallel with
  an Arco 469 variable or three or more smaller value
  mica capacitors in parallel

C12—not used
C15—10 μF, 100 V electrolytic
C16—1000 pF ceramic
C17,C18—two 0.1 μF, 100 V ceramic each,
  (ATC 200/823 or equivalent)
D1-D4—IN4148
D5—28 V zener, IN5362 or equivalent
L1,L2—Two Fair-Rite 2673021801 ferrite beads
  each or equivalent 4.0 μH
T1-T3—see text
Q1-Q4—MRF150
IC1—MC1723CP
  All resistors are 0.5W carbon or metal film
  unless otherwise designated.

Fig. 8-15

This is a unique push-pull parallel circuit. It uses four MRF150 RF power FETs paralleled at relatively high power levels. Supply voltages of 40 to 50 Vdc can be used, depending on linearity requirements. The bias for each device is independently adjustable; therefore, no matching is required for the gate threshold voltages.

6-METER KILOWATT AMPLIFIER

The amplifier uses a grounded-grid circuit with either the Eimac 3CX1000A7 or 8877 ceramic/metal triodes, which are intended for linear service in the HF and VHF ranges. The amp provides the legal power output of 1500 watts PEP and CW service with no effort and requires a driver delivering between 50 and 80 watts at 50 MHz. With a plate voltage of 3000 volts at 0.8 amps, the amplifier performs at 60 percent efficiency. The grid is grounded by means of the grid ring of the 3CX1000A7 socket, which provides a low-inductance path to ground. The amplifier is completely stable.

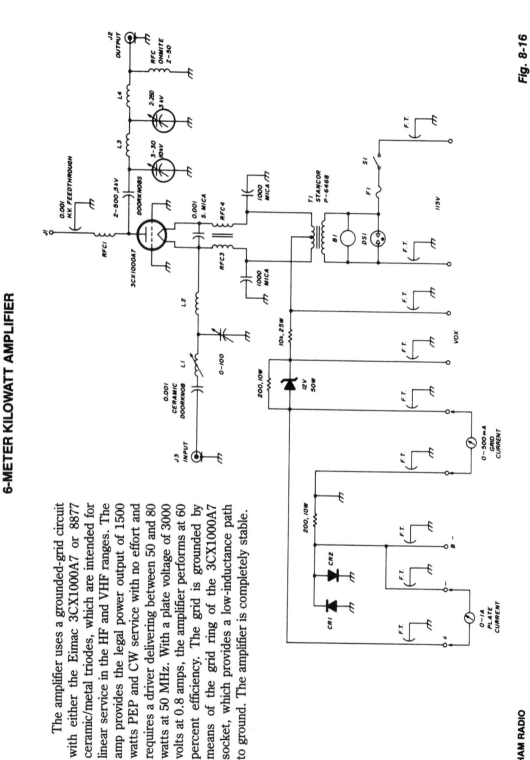

Fig. 8-16

10-dB-GAIN AMPLIFIER

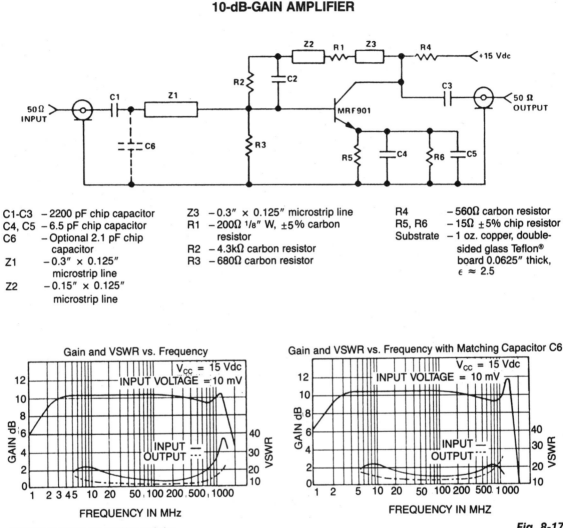

C1-C3	– 2200 pF chip capacitor	Z3	– 0.3″ × 0.125″ microstrip line	R4	– 560Ω carbon resistor
C4, C5	– 6.5 pF chip capacitor	R1	– 200Ω 1/8″ W, ±5% carbon	R5, R6	– 15Ω ±5% chip resistor
C6	– Optional 2.1 pF chip capacitor		resistor	Substrate	– 1 oz. copper, double-sided glass Teflon® board 0.0625″ thick, $\epsilon \approx 2.5$
Z1	– 0.3″ × 0.125″ microstrip line	R2	– 4.3kΩ carbon resistor		
Z2	– 0.15″ × 0.125″ microstrip line	R3	– 680Ω carbon resistor		

Fig. 8-17

This circuit design is a class A amplifier employing both ac and dc feedback. Bias is stabilized at 15 mA of the collector current using dc feedback from the collector. The ac feedback, from collector to base, and in each of the partially bypassed emitter circuits, compensates for the increase in device gain with decreasing frequency, yielding a flat response over a maximum bandwidth. The amplifier shows a nominal 10-dB power gain from 3 MHz to 1.4 GHz. With only a minimum matching network used at the amplifier input, the input VSWR remains less than 2.5:1 to approximately 1 GHz, while the output VSWR stays under 2:1. Note that a slight degradation in gain flatness and output VSWR occurs with the addition of C6. A more elaborate network design would probably optimize impedance matching, while maintaining gain flatness.

60-MHz AMPLIFIER

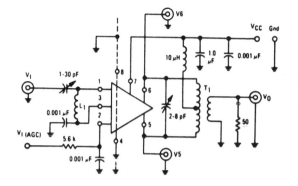

L1: = 7 Turns, #22 AWG Wire
on 5/16" Dia. Form,
5/8" Long
T1: = Close Wound Over 1/4" Form
Primary Winding = 16 Turns #26 AWG, Center Tapped
Secondary Winding = 2 Turns #26 AWG.

MOTOROLA

Fig. 8-18

30-MHz AMPLIFIER (POWER GAIN = 50 dB, BW ≈ 1.0 MHz)

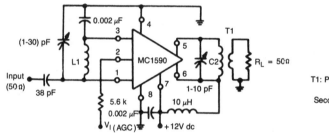

L1 = 12 Turns #22 AWG Wire on a Toroid Core.
(T37-6 Micro Metal
or Equiv.)
T1: Primary = 17b Turns #20 AWG Wire on a Toroid Core.
(T44-6 Micro Metal or Equiv)
Secondary = 2 Turns #20 AWG Wire

MOTOROLA

Fig. 8-19

2-METER AMPLIFIER (5-W OUTPUT)

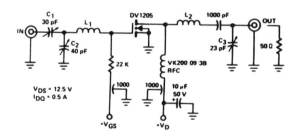

L_1, 60 nHy 4T #22 AWG close wound 0.125" I.D.
L_2, 54 nHy 31/2T #22 AWG close wound 0.125" I.D.
C_1, C_2, C_3, ARCO #462 5-80 pF

SILICONIX

Fig. 8-20

144

80-MHz CASCODE AMPLIFIER

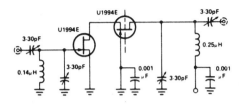

Fig. 8-21

200-MHz NEUTRALIZED COMMON-SOURCE AMPLIFIER

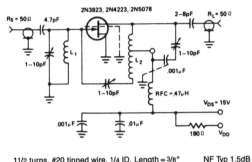

L$_1$	11/2 turns, #20 tinned wire. 1/4 ID, Length = 3/8"	NF Typ 1.5dB
L$_2$	31/2 turns, #18 tinned wire. 3/8 ID, Length = 1/2"	G$_{ps}$ Typ 18dB
	Tapped at 1 1/4 turns from drain	V$_{DS}$ = + 15V
		V$_{GS}$ = 0

TELEDYNE

Fig. 8-22

450-MHz COMMON SOURCE AMPLIFIER

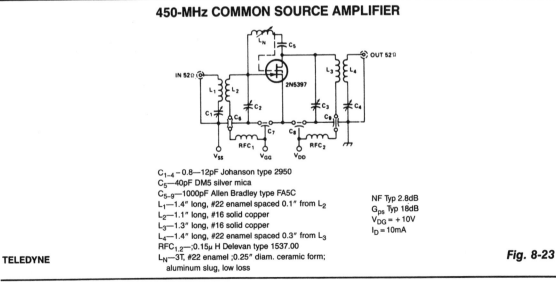

C$_{1-4}$ – 0.8 – 12pF Johanson type 2950
C$_5$—40pF DM5 silver mica
C$_{5-9}$—1000pF Allen Bradley type FA5C
L$_1$—1.4" long, #22 enamel spaced 0.1" from L$_2$
L$_2$—1.1" long, #16 solid copper
L$_3$—1.3" long, #16 solid copper
L$_4$—1.4" long, #22 enamel spaced 0.3" from L$_3$
RFC$_{1,2}$—;0.15μ H Delevan type 1537.00
L$_N$—3T, #22 enamel ;0.25" diam. ceramic form;
   aluminum slug, low loss

NF Typ 2.8dB
G$_{ps}$ Typ 18dB
V$_{DG}$ = + 10V
I$_D$ = 10mA

TELEDYNE

Fig. 8-23

1296-MHz SOLID-STATE POWER AMPLIFIER

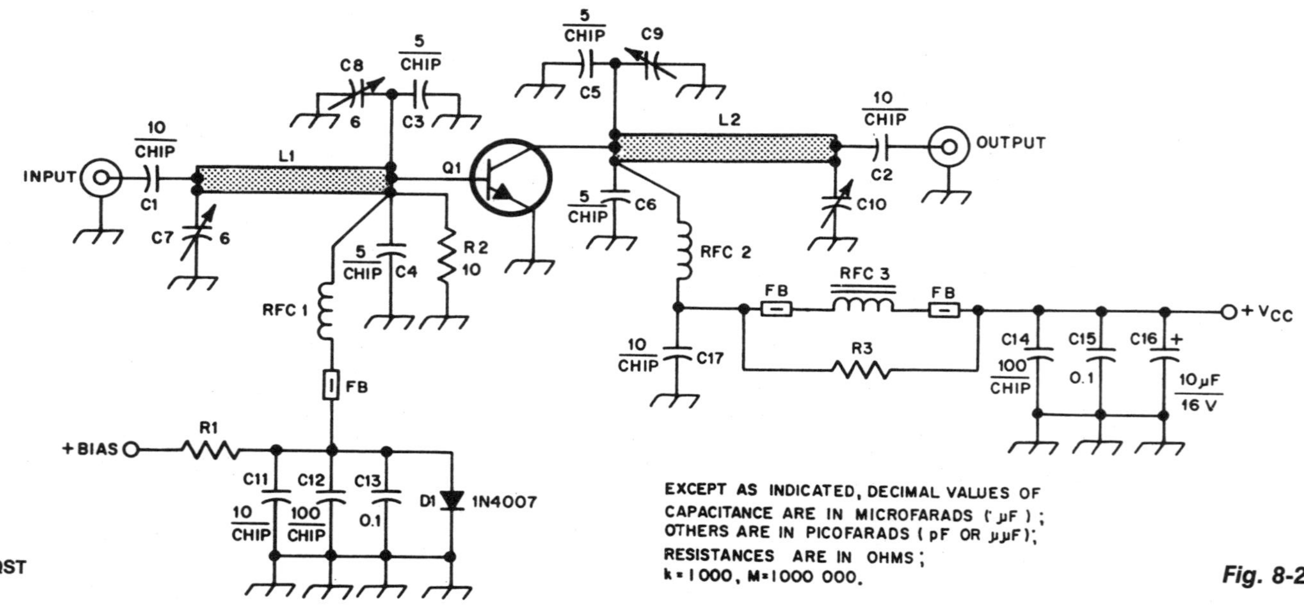

Fig. 8-24

QST

Fig. 1—Schematic diagram of the NEL1306 and NEL1320 1296-MHz solid-state power amplifiers. The schematic is identical for both versions. Component values are the same except as noted.

C1, C2, C11, C17—10-pF chip capacitor.
C3, C4, C5, C6—3.6- to 5.0-pF chip capacitor.
C7, C8—1.8- to 6.0-pF miniature trimmer capacitor (Mouser 24AA070 or equiv. See text).
C9, C10—Same as C7 and C8 for the NEL1306 amplifier. For the NEL1320 version, 0.8- to 10-pF piston trimmers are used (Johanson 5200 series or equiv.).
C12, C14—100-pF chip capacitor.

C13, C15—0.1-μF disc ceramic capacitor.
C16—10-μF electrolytic capacitor.
D1—1N4007 diode.
L1, L2—30-ohm microstripline, ¼-wavelength long (see text).
Q1—NEC NEL130681-12 (6 W) or NEL132081-12 (18 W) transistor.
R1—82- to 100-Ω resistor, 2-W minimum. Vary for specified idling current.

R2—10-Ω, ¼-W carbon-composition resistor with "zero" lead length. See text.
R3—15-Ω, 1-W carbon-composition resistor.
RFC1—3t no. 24 wire, 0.125 inch ID, spaced 1 wire diam.
RFC2—1t no. 24 wire, 0.125 inch ID, spaced 1 wire diam.
RFC3—1-μH RF choke; 18t no. 24 enam. close-spaced on a T50-10 toroid core.

EXCEPT AS INDICATED, DECIMAL VALUES OF CAPACITANCE ARE IN MICROFARADS (μF) ; OTHERS ARE IN PICOFARADS (pF OR μμF) ; RESISTANCES ARE IN OHMS ; k = 1000 , M = 1000 000.

The design incorporates 30-Ω, ¼λ microstrip lines on the input and output. C3, C4, C7, and C8, along with L1, form a pi network that matches the low-input impedance of the device to 50 Ω. C5, C6, C9, C10, and 30-Ω transmission fine L2 form an output pi network that maximizes power transfer to 50 Ω. C10 is not always necessary, depending on variations among devices and circuit-board material. Bias is provided by R1, R2, and D1. R1 can be optimized, if desired, to adjust the collector idling current.

2- to 30-MHz AMPLIFIER

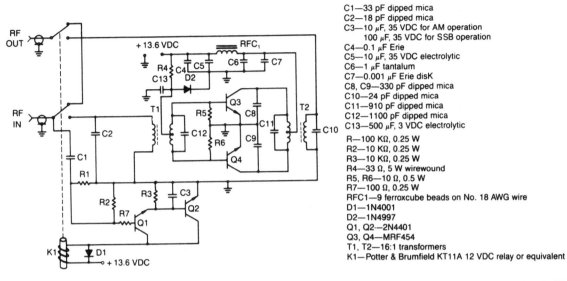

C1—33 pF dipped mica
C2—18 pF dipped mica
C3—10 µF, 35 VDC for AM operation
    100 µF, 35 VDC for SSB operation
C4—0.1 µF Erie
C5—10 µF, 35 VDC electrolytic
C6—1 µF tantalum
C7—0.001 µF Erie disK
C8, C9—330 pF dipped mica
C10—24 pF dipped mica
C11—910 pF dipped mica
C12—1100 pF dipped mica
C13—500 µF, 3 VDC electrolytic

R—100 KΩ, 0.25 W
R2—10 KΩ, 0.25 W
R3—10 KΩ, 0.25 W
R4—33 Ω, 5 W wirewound
R5, R6—10 Ω, 0.5 W
R7—100 Ω, 0.25 W
RFC1—9 ferroxcube beads on No. 18 AWG wire
D1—1N4001
D2—1N4997
Q1, Q2—2N4401
Q3, Q4—MRF454
T1, T2—16:1 transformers
K1—Potter & Brumfield KT11A 12 VDC relay or equivalent

MICROWAVES & RF

Fig. 8-25

This amplifier provides 140-W PEP nominal output power when supplied with input levels as low as 3 W. Both input and output transformers have a 4:1 turn ratio and a 16:1 impedance ratio to achieve low input VSWR across the band with high-saturation capability.

450-MHz COMMON-GATE AMPLIFIER

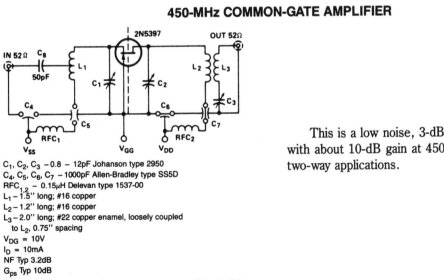

C_1, C_2, C_3 – 0.8 – 12pF Johanson type 2950
C_4, C_5, C_6, C_7 – 1000pF Allen-Bradley type SS5D
$RFC_{1,2}$ – 0.15µH Delevan type 1537-00
L_1 – 1.5'' long; #16 copper
L_2 – 1.2'' long; #16 copper
L_3 – 2.0'' long; #22 copper enamel, loosely coupled
    to L_2, 0.75'' spacing
V_{DG} = 10V
I_D = 10mA
NF Typ 3.2dB
G_{ps} Typ 10dB

This is a low noise, 3-dB typical NF, amplifier with about 10-dB gain at 450 – 470 MHz for VHF two-way applications.

Fig. 8-26

200-MHz CASCODE AMPLIFIER

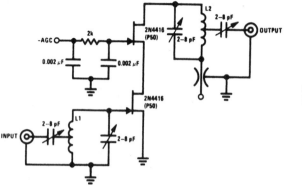

AGC range 59 dB
power gain 17 dB

L1 = 0.07 μHy center tap
L2 = 0.07 μHy tap 1/4 up from ground

NATIONAL SEMICONDUCTOR

Fig. 8-27

This 200-MHz JFET cascode circuit features low cross-modulation, large signal handling ability, no neutralization, and AGC that is controlled by biasing the upper cascode JFET. The only special requirement of this circuit is that I_{DSS} of the upper unit must be greater than that of the lower unit.

135- to 175-MHz AMPLIFIER

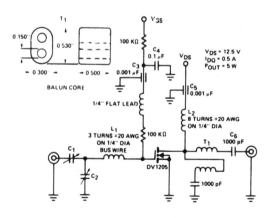

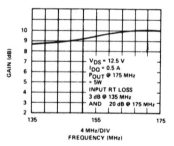

C_1, C_2 ARCO #462, 2 to 80 pF, trimmer capacitors
L_1, 3 turns buss wire #20 AWG on 1/4" diameter
L_2, 8 turns #20 AWG on 1/4" diameter
T_1, 1 turn of 25 Ω coax on 2 balun cores.
Stackpole #57-0973 μo = 35.

SILICONIX

Fig. 8-28

LOW-DISTORTION 1.6- TO 30-MHz SSB DRIVER

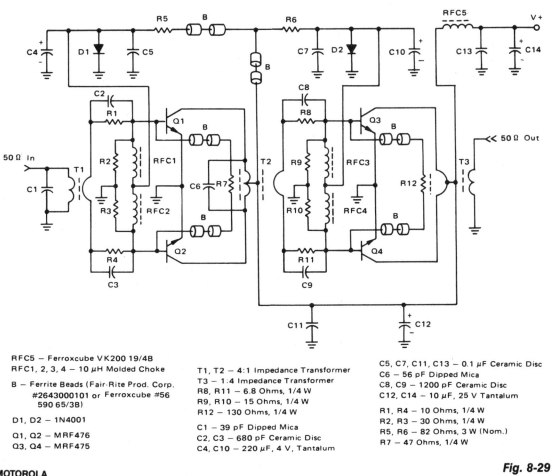

RFC5 — Ferroxcube VK200 19/4B
RFC1, 2, 3, 4 — 10 μH Molded Choke

B — Ferrite Beads (Fair-Rite Prod. Corp.
    #2643000101 or Ferroxcube #56
    590 65/3B)

D1, D2 — 1N4001

Q1, Q2 — MRF476
Q3, Q4 — MRF475

T1, T2 — 4:1 Impedance Transformer
T3 — 1:4 Impedance Transformer
R8, R11 — 6.8 Ohms, 1/4 W
R9, R10 — 15 Ohms, 1/4 W
R12 — 130 Ohms, 1/4 W

C1 — 39 pF Dipped Mica
C2, C3 — 680 pF Ceramic Disc
C4, C10 — 220 μF, 4 V, Tantalum

C5, C7, C11, C13 — 0.1 μF Ceramic Disc
C6 — 56 pF Dipped Mica
C8, C9 — 1200 pF Ceramic Disc
C12, C14 — 10 μF, 25 V Tantalum

R1, R4 — 10 Ohms, 1/4 W
R2, R3 — 30 Ohms, 1/4 W
R5, R6 — 82 Ohms, 3 W (Nom.)
R7 — 47 Ohms, 1/4 W

MOTOROLA

Fig. 8-29

The amplifier provides a total power gain of about 25 dB, and the construction technique allows the use of inexpensive components throughout. The MRF476 is specified as a 3-W device and the MRF475 has an output power of 12 watts. Both are extremely tolerant to overdrive and load mismatches, even under CW conditions. Typical IMD numbers are better than −35 dB, and the power gains are 18 dB and 12 dB, respectively, at 30 MHz. The bias currents of each stage are individually adjustable with R5 and R6. Capacitors C4 and C10 function as audio-frequency bypasses to further reduce the source impedance at the frequencies of modulation. Gain leveling across the band is achieved with simple RC networks in series with the bases, in conjunction with negative feedback. The amplitude of the out-of-phase voltages at the bases is inversely proportional to the frequency as a result of the series inductance in the feedback loop and the increasing input impedance of the transistor at low frequencies. Conversely, the negative feedback lowers the effective input impedance that is presented to the source (not the input impedance of the device itself) and with proper voltage slope, would equalize it. With this technique, it is possible to maintain an input VSWR of 1.5:1 or less than 1.6 to 30 MHz.

100- AND 400-MHz NEUTRALIZED COMMON-SOURCE AMPLIFIER

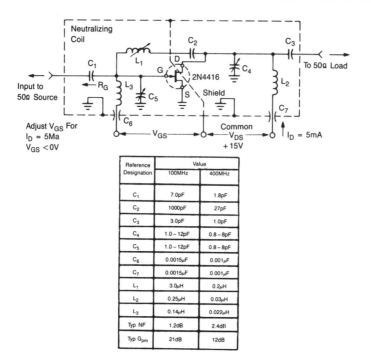

Reference Designation	Value	
	100MHz	400MHz
C_1	7.0pF	1.8pF
C_2	1000pF	27pF
C_3	3.0pF	1.0pF
C_4	1.0 – 12pF	0.8 – 8pF
C_5	1.0 – 12pF	0.8 – 8pF
C_6	0.0015μF	0.001μF
C_7	0.0015μF	0.001μF
L_1	3.0μH	0.2μH
L_2	0.25μH	0.03μH
L_3	0.14μH	0.022μH
Typ NF	1.2dB	2.4dB
Typ G_{pm}	21dB	12dB

TELEDYNE

Fig. 8-30

ULTRA-HIGH-FREQUENCY AMPLIFIER

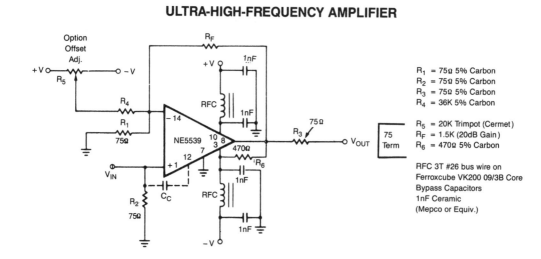

R_1 = 75Ω 5% Carbon
R_2 = 75Ω 5% Carbon
R_3 = 75Ω 5% Carbon
R_4 = 36K 5% Carbon

R_5 = 20K Trimpot (Cermet)
R_F = 1.5K (20dB Gain)
R_6 = 470Ω 5% Carbon

RFC 3T #26 bus wire on
Ferroxcube VK200 09/3B Core
Bypass Capacitors
1nF Ceramic
(Mepco or Equiv.)

SIGNETICS

Fig. 8-31

UHF AMPLIFIER

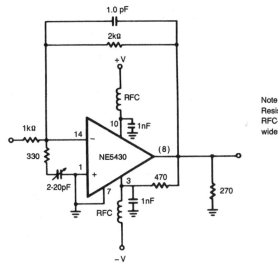

Note
Resistors – 1/4 watt carbon.
RFC-3T #26 bus wire on Ferroxcube VK200 09/3B
wideband threaded core.

SIGNETICS

Fig. 8-32

This UHF amplifier has an inverting gain of 2 and lag-lead compensation. The gain bandwidth product is 350 MHz.

TRANSISTORIZED Q-MULTIPLIER

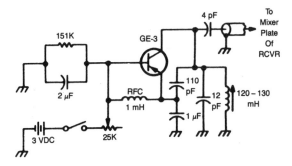

73 AMATEUR RADIO

Fig. 8-33

This transistorized Q-multiplier is for use with IFs in the 1400-kHz range.

6-METER PREAMPLIFIER PROVIDES 20-dB GAIN AND LOW NF

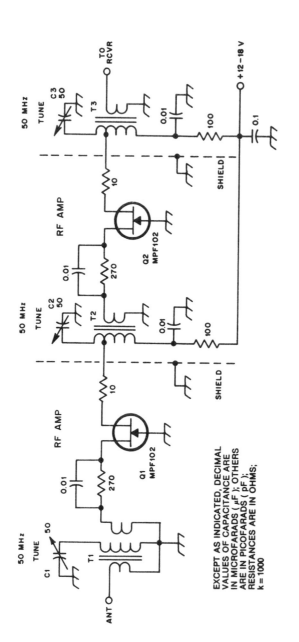

EXCEPT AS INDICATED, DECIMAL VALUES OF CAPACITANCE ARE IN MICROFARADS (μF); OTHERS ARE IN PICOFARADS (pF); RESISTANCES ARE IN OHMS; k = 1000

Fig. 8-34

QST

C1, C2, and C3 are miniature ceramic or plastic trimmers. T1 (main winding) is 0.34 μH. Use of 11 turns of #24 enameled wire on a T37-10 toroid core. The antenna winding has one turn and Q1 the source winding has three turns. T2 primary consists of 11 turns of #24 enameled wire on a T37-10 toroid. Tap Q1 drain is three turns from C2 the end of the winding. The secondary has three turns. T3 is the same as T2, except its secondary has one turn.

28-dB NONINVERTING AMPLIFIER

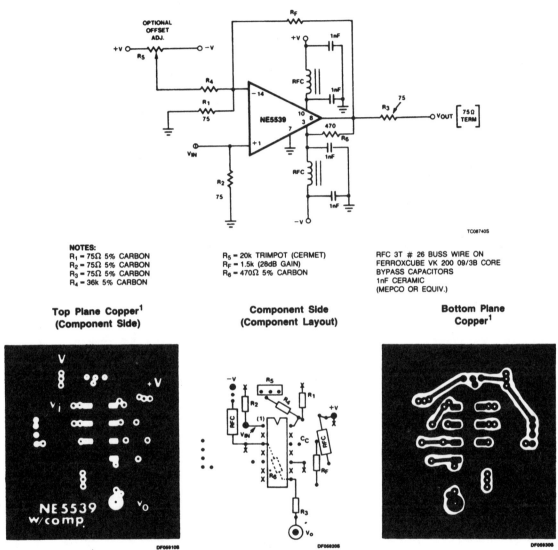

NOTES:
$R_1 = 75\Omega$ 5% CARBON
$R_2 = 75\Omega$ 5% CARBON
$R_3 = 75\Omega$ 5% CARBON
$R_4 = 36k$ 5% CARBON

$R_5 = 20k$ TRIMPOT (CERMET)
$R_F = 1.5k$ (28dB GAIN)
$R_6 = 470\Omega$ 5% CARBON

RFC 3T # 26 BUSS WIRE ON
FERROXCUBE VK 200 09/3B CORE
BYPASS CAPACITORS
1nF CERAMIC
(MEPCO OR EQUIV.)

Top Plane Copper[1]
(Component Side)

Component Side
(Component Layout)

Bottom Plane
Copper[1]

NOTES:
(X) indicates ground connection to top plane.
*R_6 is on bottom side.

NOTE:
1. Bond edges of top and bottom ground plane copper.

SIGNETICS

Fig. 8-35

The physical circuit layout is extremely critical. Breadboarding is not recommended. A double-sided copper-clad printed circuit board will result in more favorable system operation.

IMPROVED RF ISOLATION AMPLIFIER

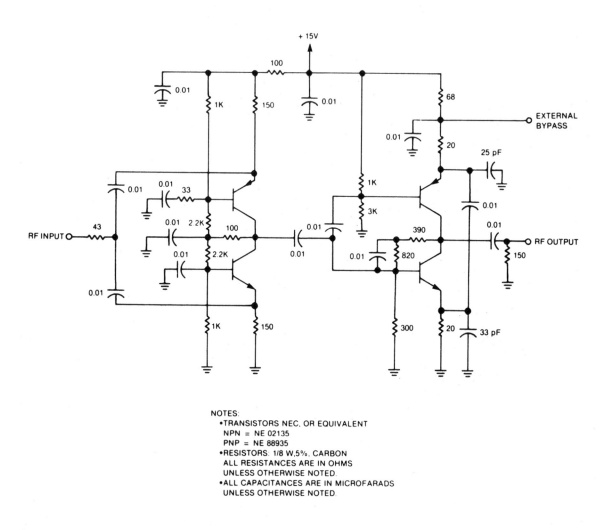

NOTES:
- TRANSISTORS NEC, OR EQUIVALENT
  NPN = NE 02135
  PNP = NE 88935
- RESISTORS: 1/8 W,5%. CARBON
  ALL RESISTANCES ARE IN OHMS
  UNLESS OTHERWISE NOTED.
- ALL CAPACITANCES ARE IN MICROFARADS
  UNLESS OTHERWISE NOTED.

NASA

Fig. 8-36

This wideband RF isolation amplifier has a frequency response of 0.5 to 400 MHz ± 0.5 dB. This two-stage amplifier can be used in applications requiring high reverse isolation, such as receiver intermediate-frequency (IF) strips and frequency-distribution systems. Both stages use complementary-symmetry transistor arrangements. The input stage is a common-base connection for the complementary circuit. The output stage, which supplies the positive gain, is a common-emitter circuit using emitter degeneration and collector-base feedback for impedance control.

2-STAGE 60-MHz IF AMPLIFIER
(POWER GAIN ≈ 80 dB, BW ≈ 1.5 MHz)

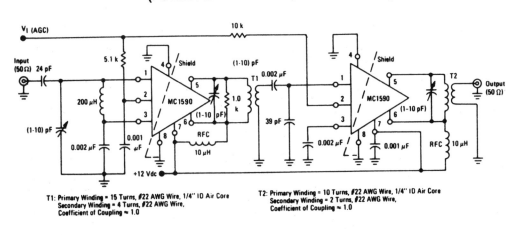

T1: Primary Winding = 15 Turns, #22 AWG Wire, 1/4" ID Air Core
Secondary Winding = 4 Turns, #22 AWG Wire,
Coefficient of Coupling ≈ 1.0

T2: Primary Winding = 10 Turns, #22 AWG Wire, 1/4" ID Air Core
Secondary Winding = 2 Turns, #22 AWG Wire,
Coefficient of Coupling ≈ 1.0

MOTOROLA

Fig. 8-37

28-V WIDEBAND AMPLIFIER (3 to 100 MHz)

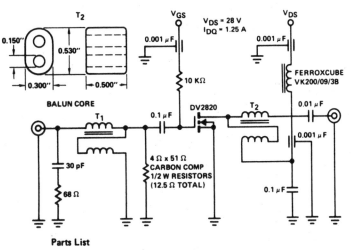

Parts List

T_1, 20 turns 30 Ω, #30 bifilar on micrometals T-50-6 Toroid
T_2, 1 turn of 2-50 Ω coax cables in parallel through 2 balun
cores stackpole #57-9130 μ_o = 125

SILICONIX

Fig. 8-38

29-MHz AMPLIFIER

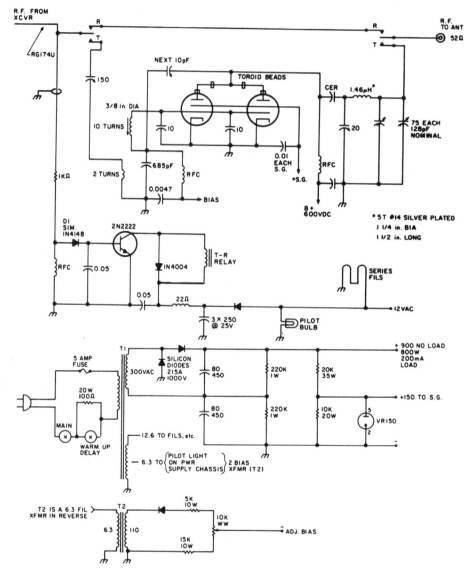

73 AMATEUR RADIO

Fig. 8-39

The only adjustments that require close attention are input, output, and neutralization. The 150-pF capacitor in the input line compensates for impedance mismatch. You tune for maximum signal transfer from exiter to final with an in-line meter or external field strength meter. The final is a conventional pi network. When neutralized, the plate current dip should be at about the same setting of the 20-pF plate capacitor as maximum output. Adjust bias to let tubes idle at about 30 mA.

RF WIDEBAND ADJUSTABLE AGC AMPLIFIER

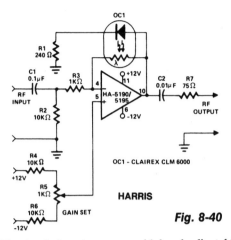

OC1 - CLAIREX CLM 6000

HARRIS

Fig. 8-40

This circuit functions as a wideband adjustable AGC amplifier. With an effective bandwidth of approximately 10 MHz, it is capable of handling rf input signal frequencies from 3.2 to 10 MHz at levels ranging from 40 mV up to 3 V pk-pk.

AGC action is achieved by using optocoupler/isolater OC1 as part of the gain control-feedback loop. In operation, the positive peaks of the amplified output signal drive the OC1 LED into a conducting state. Because the resistance of the OC1 photosensitive element is inversely proportional to light intensity, the higher the signal level, the lower the feedback resistance to the op amp inverting input. The greater negative feedback lowers stage gain. Any changes in gain occur smoothly because the inherent memory characteristic of the photoresistor acts to integrate the peak signal inputs. In practice, the stage gain is adjusted automatically to where the output signal positive peaks are approximately one diode drop above ground.

Gain set control R5 applies a fixed dc bias to the op amp noninverting input, thus establishing the steady state-zero input signal current through the OC1 LED and determining the signal level at which AGC action begins.

The effective AGC range depends on a number of factors, including individual device characteristics, the nature of the rf drive signal, the initial setting for R5, et al. Theoretically, the AGC range can be as high as 4000:1 for a perfect op amp because the OC1 photoresistor can vary in value from 1 MΩ with the LED dark to 250 Ω with the LED fully on.

1-MHz METER-DRIVER AMPLIFIER

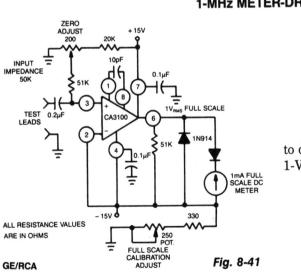

ALL RESISTANCE VALUES ARE IN OHMS

GE/RCA

Fig. 8-41

This circuit uses the CA3100 BiMOS op amp to drive a 1-mA meter movement to full scale with 1-V rms input.

WIDEBAND UHF AMPLIFIER WITH HIGH-PERFORMANCE FETS

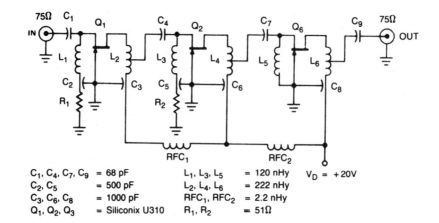

C_1, C_4, C_7, C_9	= 68 pF	L_1, L_3, L_5	= 120 nHy	V_D = +20V
C_2, C_5	= 500 pF	L_2, L_4, L_6	= 222 nHy	
C_3, C_6, C_8	= 1000 pF	RFC_1, RFC_2	= 2.2 nHy	
Q_1, Q_2, Q_3	= Siliconix U310	R_1, R_2	= 51Ω	

SILICONIX

Fig. 8-42

The amplifier circuit is designed for a 225 MHz center frequency, 1 dB bandwidth of 50 MHz, low-input VSWR in a 75-Ω system, and 24 dB gain. Three stages of U310 FETs are used, in a straight-forward design.

BROADCAST-BAND RF AMPLIFIER

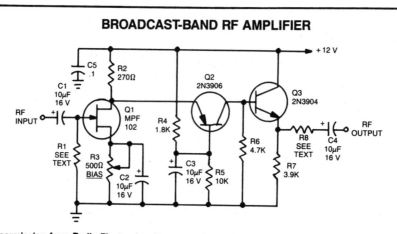

Reprinted with permission from Radio-Electronics Magazine, 1989 R-E Experimenters Handbook. Copyright Gernsback Publications, Inc., 1989.

Fig. 8-43

The circuit has a frequency response ranging from 100 Hz to 3 MHz; gain is about 30 dB. Field-effect transistor Q1 is configured in the common-source self-biased mode. Optional resistor R1 allows you to set the input impedance to any desired value; commonly, it will be 50 Ω.

The signal is then direct coupled to Q2, a common-base circuit that isolates the input and output stages and provides the amplifier's exceptional stability. Last, Q3 functions as an emitter follower, to provide low output impedance at about 50 Ω. If you need higher output impedance, include resistor R8. It will affect impedance according to this formula: $R8 \approx R_{OUT} - 50$. Otherwise, connect output capacitor C4 directly to the emitter of Q3.

MINIATURE WIDEBAND AMPLIFIER

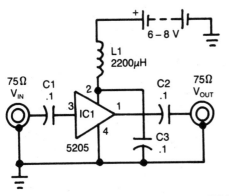

This wideband amplifier uses only five components. External signals enter pin 3 of IC1 via ac coupling capacitor C1. Following amplification, the boosted signals from IC1 pin 1 are coupled to the output by capacitor C2. Capacitor C3 decouples the dc power supply, while rf current is isolated from the power supply by rf choke L1.

The NE5205's low current consumption of 25 mA at 6 Vdc makes battery-powered operation a reality. Although the device is rated for a 6 to 8 V power supply, 6 V is recommended for normal operation. From 6 V an internal bias of 3.3 V results, which permits a 1.4 V pk-pk output swing for video applications.

Fig. 8-44

WIDEBAND 500-kHz to 1-GHz HYBRID AMPLIFIER

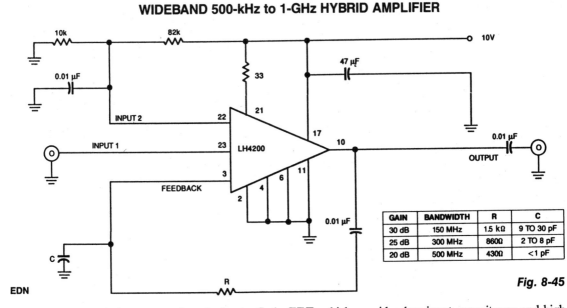

GAIN	BANDWIDTH	R	C
30 dB	150 MHz	1.5 kΩ	9 TO 30 pF
25 dB	300 MHz	860Ω	2 TO 8 pF
20 dB	500 MHz	430Ω	<1 pF

EDN

Fig. 8-45

The amplifier's input stage is a dual-gate GaAs FET, which provides low input capacitance and high transconductance. The dual-gate structure accepts the signal on input 1. Input 2 controls the gain of the amplifier. The amplifier has a third input for use in series feedback. The output feeds back to pin 3 via a single resistor, which controls the overall power gain of the amplifier. At 10 MHz, the output is capable of delivering 12 dBm into a 50-Ω load with 1 dB of signal compression. The ac-coupled amplifier has a gain of 37 dB at 100 MHz and 3 dB at 1 GHz.

SHORTWAVE FET BOOSTER

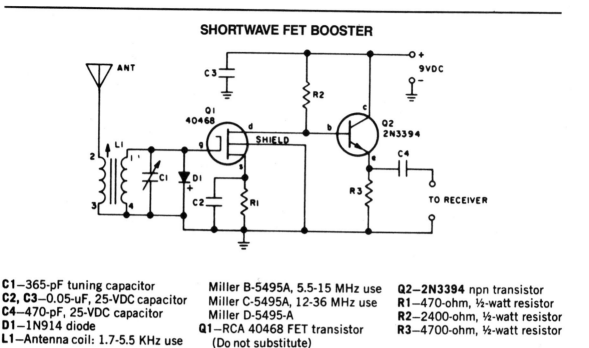

C1—365-pF tuning capacitor
C2, C3—0.05-uF, 25-VDC capacitor
C4—470-pF, 25-VDC capacitor
D1—1N914 diode
L1—Antenna coil: 1.7-5.5 KHz use

Miller B-5495A, 5.5-15 MHz use
Miller C-5495A, 12-36 MHz use
Miller D-5495-A
Q1—RCA 40468 FET transistor
(Do not substitute)

Q2—2N3394 npn transistor
R1—470-ohm, ½-watt resistor
R2—2400-ohm, ½-watt resistor
R3—4700-ohm, ½-watt resistor

EDN

Fig. 8-46

This two transistor preselector provides up to 40-dB gain from 3.5 to 30 MHz. Q1 (a MOSFET) is sensitive to static charges and must be handled with care.

LOW-NOISE 30-MHz PREAMPLIFIER

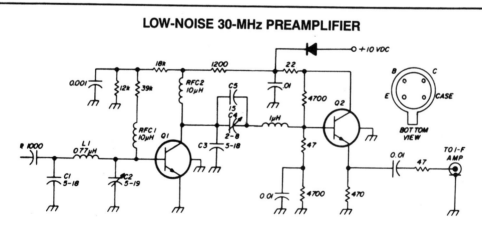

HANDS-ON ELECTRONICS

Fig. 8-47

Low-noise preamplifier has a noise figure of 1.1 dB at 30 MHz and 3-dB bandwidth of 10 MHz. Gain is 19 dB. Total current drain with a +10-V supply is 13 mA. All resistors are ¼-W carbon; bypass capacitors are 50-V ceramics.

WIDEBAND UHF AMPLIFIER WITH HIGH-PERFORMANCE FETs

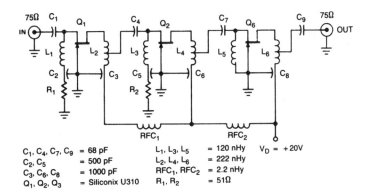

C_1, C_4, C_7, C_9	= 68 pF	L_1, L_3, L_5	= 120 nHy
C_2, C_5	= 500 pF	L_2, L_4, L_6	= 222 nHy
C_3, C_6, C_8	= 1000 pF	RFC_1, RFC_2	= 2.2 nHy
Q_1, Q_2, Q_3	= Siliconix U310	R_1, R_2	= 51Ω

$V_D = +20V$

MAXIM

Fig. 8-48

The amplifier circuit is designed for 225-MHz center frequency, 1-dB bandwidth of 50 MHz, low-input VSWR in a 75-Ω system, and 24-dB gain. Three stages of U310 FETs are used in a straight-forward design.

10-MHz COAXIAL LINE DRIVER

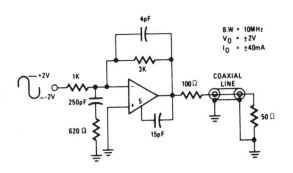

The circuit will find excellent usage in high-frequency line-driving systems that require wide-power bandwidths at high output-current levels. (IC = HA2530) The bandwidth of the circuit is limited only by the single-pole response of the feedback components; namely $f(-3 \text{ dB}) = 1/2 \pi R_f C_f$. As such, the response is flat with no peaking and yields minimum distortion.

TEXAS INSTRUMENTS

Fig. 8-49

VHF PREAMPLIFIER

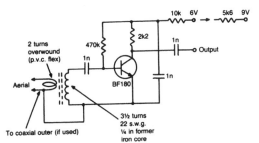

This simple circuit gives 15-dB gain and it can be mounted on a 1-inch-square printed circuit board. Coil data is given for 85 to 95 MHz. For other frequencies, modify the coil, as required.

TEXAS INSTRUMENTS

Fig. 8-50

9

Transducer Amplifiers

The sources of the following circuits are contained in the Sources section, which begins on page 182. The figure number contained with each circuit correlates to the source entry in the Sources section.

DIFFERENTIAL-TO-SINGLE-ENDED VOLTAGE AMPLIFIER

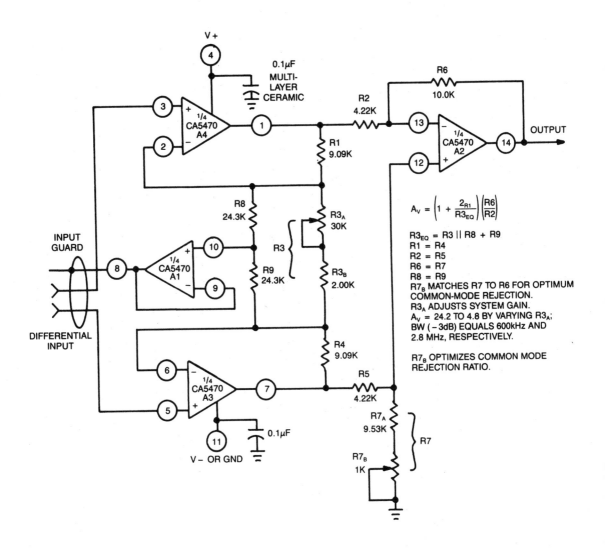

$$A_V = \left(1 + \frac{2_{R1}}{R3_{EQ}}\right)\left(\frac{R6}{R2}\right)$$

$R3_{EQ} = R3 \parallel R8 + R9$
$R1 = R4$
$R2 = R5$
$R6 = R7$
$R8 = R9$
$R7_B$ MATCHES R7 TO R6 FOR OPTIMUM COMMON-MODE REJECTION.
$R3_A$ ADJUSTS SYSTEM GAIN.
A_V = 24.2 TO 4.8 BY VARYING $R3_A$;
BW (–3dB) EQUALS 600kHz AND
2.8 MHz, RESPECTIVELY.

$R7_B$ OPTIMIZES COMMON MODE
REJECTION RATIO.

GE/RCA

Fig. 9-1

This circuit uses a CA5470 quad microprocessor BiMOS-E op amp. Amplifiers A1 and A2 are employed as a cross-coupled differential input and differential output preamp stage and A3 provides input guard-banding. Amplifier A4 converts the differential outputs of A1 and A2 to a single-ended output.

EQUALIZED PREAMP FOR MAGNETIC PHONOGRAPH CARTRIDGES

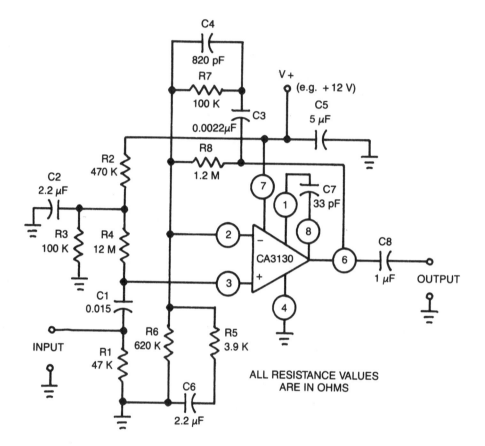

GE/RCA

Fig. 9-2

This circuit uses a CA3130 BiMOS op amp. Amplifier is equalized to RIAA playback frequency-response specifications. The circuit is useful as preamplifier following a magnetic tapehead.

PHOTODIODE AMPLIFIER

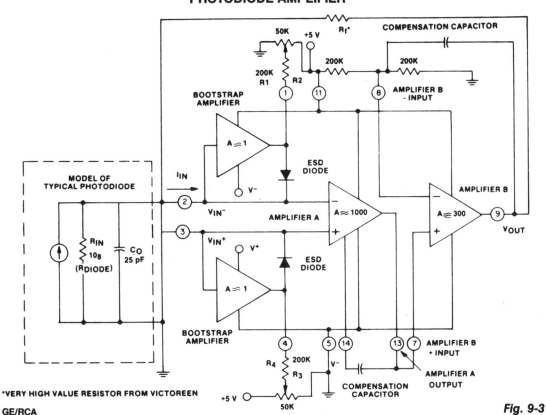

*VERY HIGH VALUE RESISTOR FROM VICTOREEN

GE/RCA

Fig. 9-3

This circuit uses a CA5422 dual BiMOS microprocessor op amp. The bootstrap amplifiers minimize bias currents while maintaining electrostatic discharge protection. Additionally, the potentiometers and their associated resistors, R1 through R4, permit the user to trim bias currents to zero.

TAPE PLAYBACK AMPLIFIER

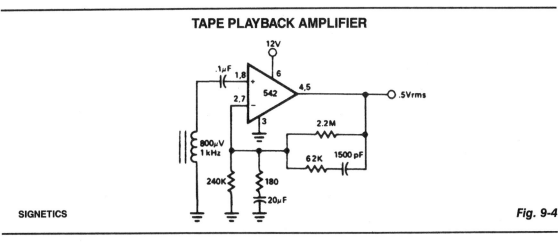

SIGNETICS

Fig. 9-4

165

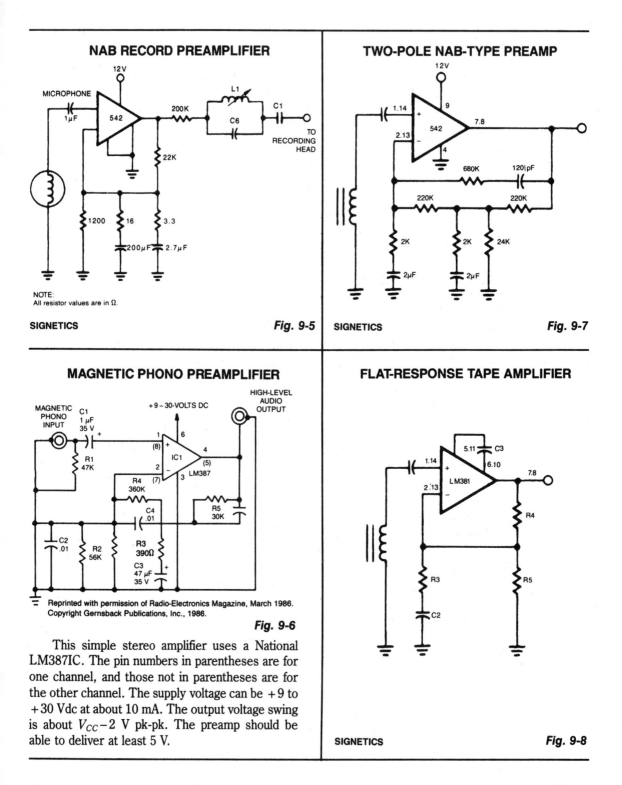

NAB RECORD PREAMPLIFIER

MICROPHONE

12 V

542

1μF

200K

L1

C6

C1

TO
RECORDING
HEAD

22K

1200   16   3.3

200μF   2.7μF

NOTE:
All resistor values are in Ω.

SIGNETICS

Fig. 9-5

TWO-POLE NAB-TYPE PREAMP

12 V

1.14

9

542

7.8

2.13

4

680K   120 pF

220K   220K

2K   2K   24K

2μF   2μF

SIGNETICS

Fig. 9-7

MAGNETIC PHONO PREAMPLIFIER

MAGNETIC
PHONO
INPUT

C1
1 μF
35 V

+9 – 30-VOLTS DC

HIGH-LEVEL
AUDIO
OUTPUT

1
(8)

6

+

IC1

4
(5)

R1
47K

2
(7)

–

3

LM387

R4
360K

R5
30K

C4
.01

C2
.01

R2
56K

R3
390Ω

C3
47 μF
35 V

Reprinted with permission of Radio-Electronics Magazine, March 1986.
Copyright Gernsback Publications, Inc., 1986.

Fig. 9-6

This simple stereo amplifier uses a National
LM387IC. The pin numbers in parentheses are for
one channel, and those not in parentheses are for
the other channel. The supply voltage can be +9 to
+30 Vdc at about 10 mA. The output voltage swing
is about $V_{CC} - 2$ V pk-pk. The preamp should be
able to deliver at least 5 V.

FLAT-RESPONSE TAPE AMPLIFIER

5.11   C3

1.14

6.10

+

LM381

7.8

2.13

–

R4

R3

R5

C2

SIGNETICS

Fig. 9-8

166

LOW-NOISE PHOTODIODE AMPLIFIER

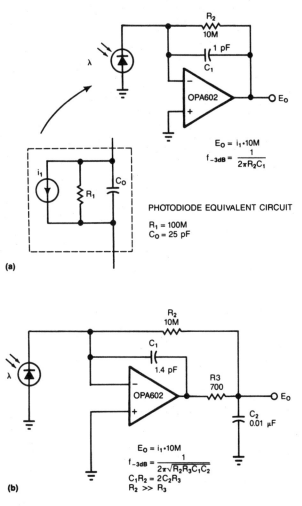

$E_O = i_1 \cdot 10M$

$f_{-3dB} = \dfrac{1}{2\pi R_2 C_1}$

PHOTODIODE EQUIVALENT CIRCUIT

$R_1 = 100M$
$C_O = 25\ pF$

(a)

$E_O = i_1 \cdot 10M$

$f_{-3dB} = \dfrac{1}{2\pi\sqrt{R_2 R_3 C_1 C_2}}$

$C_1 R_2 = 2C_2 R_3$

$R_2 \gg R_3$

(b)

EDN

Fig. 9-9

Adding two passive components to a standard photodiode amplifier reduces noise. Without the modification, the shunt capacitance of the photodiode reacting with the relatively large feedback resistor of the transimpedance (current-to-voltage) amplifier, creates excessive noise gain.

The improved circuit, Fig. 9-9b, adds a second pole, formed by R3 and C2. The modifications reduce noise by a factor of 3. Because the pole is within the feedback loop, the amplifier maintains its low output impedance. If you place the pole outside the feedback loop, you have to add an additional buffer, which would increase noise and dc error.

The signal bandwidth of both circuits is 16 kHz. In the standard circuit (Fig. 9-9a), the 1-pF stray capacitance in the feedback loop forms a single 16-kHz pole. The improved circuit has the same bandwidth as the first, but exhibits a 2-pole response.

167

MAGNETIC-PICKUP PHONO PREAMPLIFIER

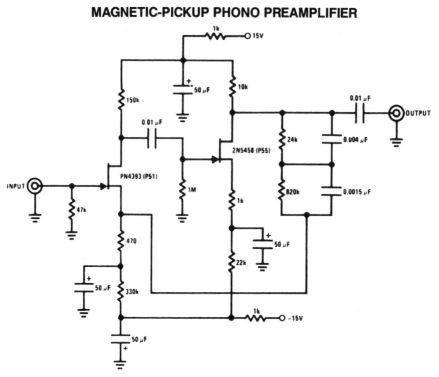

NATIONAL SEMICONDUCTOR

Fig. 9-10

This preamplifier provides proper loading to a reluctance phono cartridge. It provides approximately 35 dB of gain at 1 kHz (2.2-mV input for 100-mV output). It features (S + N)/N ratio of better than − 70 dB (referenced to 10 mV input at 1 kHz) and has a dynamic range of 84 dB (referenced to 1 kHz). The feedback provides for RIAA equalization.

DISC/TAPE PHASE-MODULATED READBACK SYSTEM

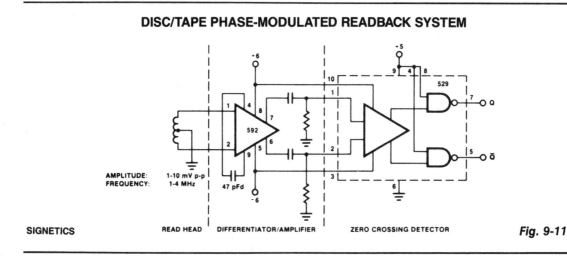

SIGNETICS

Fig. 9-11

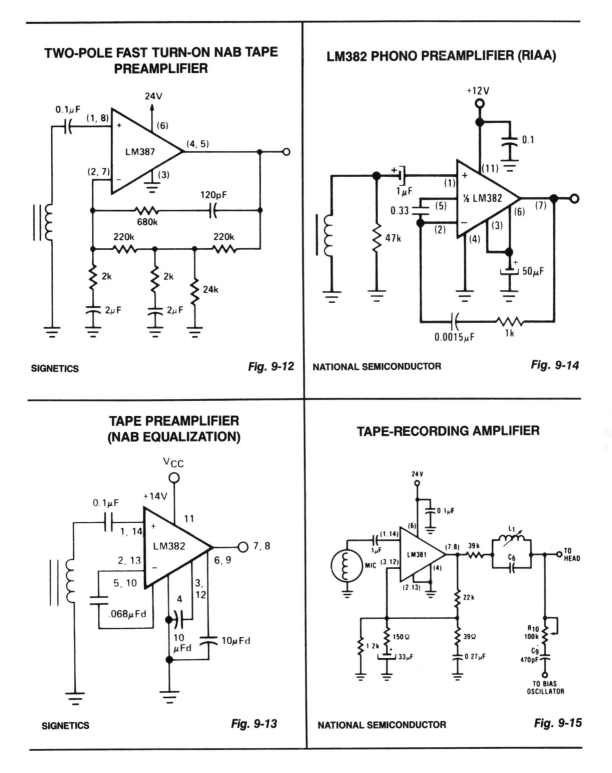

TWO-POLE FAST TURN-ON NAB TAPE PREAMPLIFIER

24V

0.1μF (1, 8) +
(6)
LM387 (4, 5)
(2, 7) −
(3)

120pF
680k
220k   220k
2k   2k   24k
2μF   2μF

SIGNETICS   *Fig. 9-12*

LM382 PHONO PREAMPLIFIER (RIAA)

+12V

0.1
(11)
(1) +
1μF   ½ LM382
0.33 (5)   (6) (7)
(2) −   (3)
47k   (4)
+ 50μF

0.0015μF   1k

NATIONAL SEMICONDUCTOR   *Fig. 9-14*

TAPE PREAMPLIFIER (NAB EQUALIZATION)

Vcc

+14V
0.1μF   1, 14 +
LM382   11
2, 13 −   7, 8
5, 10   6, 9
.068μFd   4   3, 12
10 μFd   10μFd

SIGNETICS   *Fig. 9-13*

TAPE-RECORDING AMPLIFIER

24 V

0.1μF
(1,14)
(6)
1μF
MIC (3,12)   LM381   (7,8) 39k   L₁
(4)   C₆   TO HEAD
(2,13)
22 k
1 2k   150Ω   39Ω
33μF   0 27μF   R₁₀ 100k
C₉ 470pF
TO BIAS OSCILLATOR

NATIONAL SEMICONDUCTOR   *Fig. 9-15*

169

BALANCED-INPUT MICROPHONE AMPLIFIER

It is possible to simulate the balanced performance of a transformer electronically with a differential amplifier. By adjusting the presets, the resistor ratio can be balanced so that the best CMRR is obtained. It is possible to get a better CMRR than from a transformer. Use a RC4136, quad low-noise op amp.

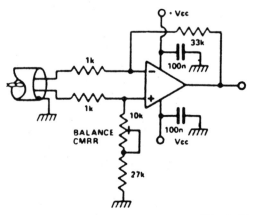

ELECTRONICS TODAY INTERNATIONAL

Fig. 9-16

TRANSDUCER AMPLIFIER

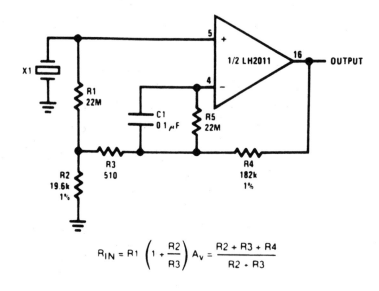

$$R_{IN} = R1 \left(1 + \frac{R2}{R3}\right) \qquad A_v = \frac{R2 + R3 + R4}{R2 + R3}$$

NATIONAL SEMICONDUCTOR

Fig. 9-17

This circuit is high-input-impedance ac amplifier for a piezoelectric transducer. Input resistance is 880 MΩ, and a gain of 10 is obtained.

10

Video Amplifiers

The sources of the following circuits are contained in the Sources section, which begins on page 182. The figure number contained with each circuit correlates to the source entry in the Sources section.

RGB VIDEO AMPLIFIER

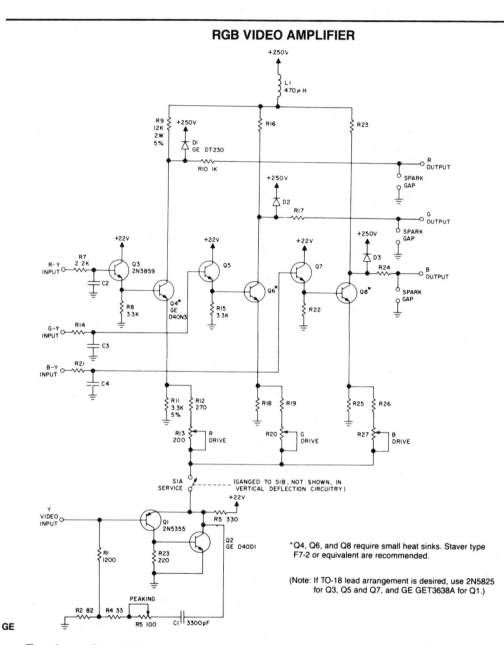

Fig. 10-1

Transistors Q1 and Q2 and their associated components provide: a low-impedance output with the necessary power to drive the output stages, give increased gain to high frequencies, and peaking the video for enhanced transient response. Emitter followers Q3, Q5, and Q7 provides low-impedance drive to output stages, Q4, Q6, and Q8. The output stages, with the color difference signals applied to their bases and the luminance signals to their emitters, perform matrixing. The matrixing results in composite output information, to the picture tube, which contains both luminance and chroma information.

VIDEO IF AMPLIFIER AND LOW-LEVEL VIDEO DETECTOR

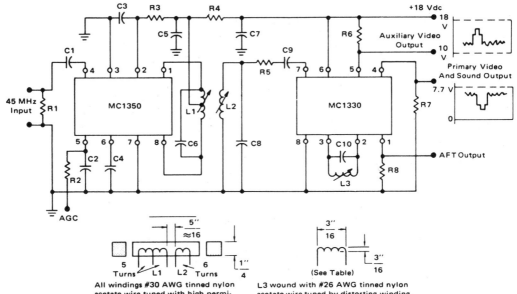

All windings #30 AWG tinned nylon
acetate wire tuned with high permiability slugs. Coil Craft #4786
differential transformer.

L3 wound with #26 AWG tinned nylon
acetate wire tuned by distorting winding.

Table of Component Values

Component	36 MHz	·45 MHz	58 MHz
C6	24 pF	15 pF	10 pF
C8	18 pF	12 pF	10 pF
C10	33 pF	33 pF	18 pF
L3	12 Turns	10 Turns	10 Turns

C1 = 0.001 μF   C6 = See Table   R1 = 50 Ω   R6 = 3.3 kΩ
C2 = 0.002 μF   C7 = 0.1 μF   R2 = 5 k   R7 = 3.9 kΩ
C3 = 0.002 μF   C8 = See Table   R3 = 470 Ω   R8 = 3.9 kΩ
C4 = 0.002 μF   C9 = 68 pF   R4 = 220 Ω   All Resistors
C5 = 0.002 μF   C10 = See Table   R5 = 22 Ω   1/4-W ±10%

All Caps Marked μF Ceramic HiK
All Caps Marked pF Silver Mica 5%

MOTOROLA

Fig. 10-2

The circuit has a typical voltage gain of 84 dB and a typical AGC range of 80 dB. It gives very small changes in bandpass shape, usually less than 1-dB tilt for 60-dB compression. No sections are shielded. The detector uses a single-tuned circuit (L3 and C10). Coupling between the two integrated circuits is achieved by a double-tuned transformer (L1 and L2).

VIDEO GAIN BLOCK

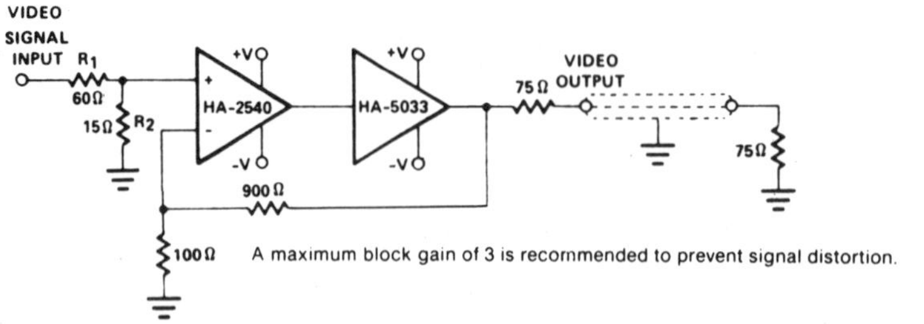

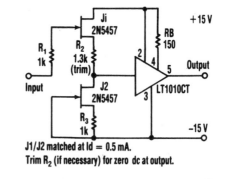

A maximum block gain of 3 is recommended to prevent signal distortion.

HARRIS

Fig. 10-3

This configuration utilizes the wide bandwidth and speed of HA-2540, plus the output capability of HA5033. Stabilization circuitry is avoided by operating HA-2540 at a closed loop gain of 10, while maintaining an overall block gain of unity. However, gain of the block can be varied using the equation:

$$\frac{V_{\text{OUT}}}{V_{\text{IN}}} = 5 \frac{R_2}{(R_1 + R_2)}$$

where $R_1 + R_2 = 75 \ \Omega$

A maximum block gain of 3 is recommended to prevent signal distortion.

This circuit was tested for differential phase and differential gain using a Tektronix 520A vector scope and a Tektronix 146 video signal generator. Both differential phase and differential gain were too small to be measured.

LOW-DISTORTION VIDEO BUFFER

This buffer amplifier's overall harmonic distortion is a low 0.01% or less at 3-V rms output into a 500-Ω load with no overall feedback. The LT1010CT offers a 100 V/μs slew rate, a 20 MHz video bandwidth, and 100 mA of output. A pair of JFETs, J1 and J2 are preselected for a nominal match at the bias level of the linearized source-follower input stage, at about 0.5 mA. The source-bias resistor, R2, of J1 is somewhat larger than R3 so that it can drop a larger voltage and cancel the LT1010CT's offset. J1 and J2 provide an untrimmed dc offset of ±50 mV or less. Swapping J1 and J2 or trimming the R2 value can give a finer match.

The circuit's overall harmonic distortion is low: 0.01% or less at 3-V rms output into a 500-Ω load with no overall feedback. The circuit's response to a ±5 V, 10 kHz square-wave input, band-limited to 1 μs, has no overshoot. If needed, setting bias

J1/J2 matched at Id = 0.5 mA.

Trim R₂ (if necessary) for zero dc at output.

ELECTRONIC DESIGN

Fig. 10-4

resistor R_B lower can accommodate even steeper input-signal slopes and drive lower impedance loads with high linearity. The main trade-off for both objectives is more power dissipation. A secondary trade-off is the need for retrimming the source-bias resistor, R2.

FET CASCODE VIDEO AMPLIFIER

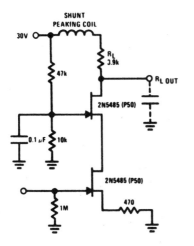

The FET cascode video amplifier features very low input loading and reduction of feedback to almost zero. The 2N5485 is used because of its low capacitance and high Y_{fs}. Bandwidth of this amplifier is limited by RL and load capacitance.

NATIONAL SEMICONDUCTOR

Fig. 10-5

HIGH-IMPEDANCE LOW-CAPACITANCE AMPLIFIER

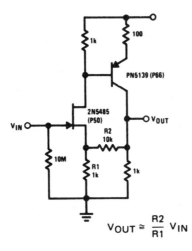

$$V_{OUT} \cong \frac{R2}{R1} V_{IN}$$

NATIONAL SEMICONDUCTOR

Fig. 10-6

This compound series-feedback circuit provides high input impedance and stable wide-band gain for general-purpose video-amplifier applications.

DC GAIN-CONTROLLED VIDEO AMPLIFIER

This amplifier employs a cascaded op amp integrator and transistor buffer, Q1, to drive the gain control element. Except for a simple modification, the HA-5190/5195 stage is connected as a conventional noninverting op amp, and includes input and output impedance matching resistors R1 and R4, respectively, series stabilization resistor R2, and power supply bypass capacitors C1 and C2. The circuit differs from standard designs in that the gain control network includes a photoresistor, part of OC1. The optocoupler/isolator OC1 contains two matched photoresistors, both activated by a common LED. The effective resistances offered by these devices are inversely proportional to the light emitted by the LED. One photoresistor is part, with R3, of the HA-5190/5195 gain network, while the other forms a voltage-divider with R6 to control the bias that is applied to the integrator noninverting terminal.

In operation, the dc voltage supplied by gain control R8 is applied to the integrator inverting input terminal through input resistor R7. Depending on the relative magnitude of the control voltage, the integrator output will either charge or discharge C3. This change in output, amplified by Q1, controls the current supplied to the OC1 LED through series limiting resistor R5. The action continues until the voltage applied to the integrator noninverting input by the R6—photoresistor gain network is

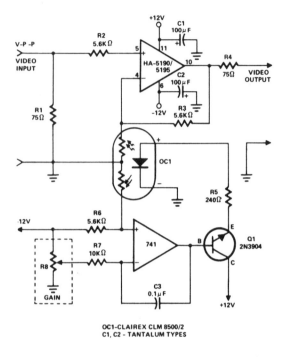

OC1-CLAIREX CLM 8500/2
C1, C2 - TANTALUM TYPES

HARRIS   *Fig. 10-7*

changing, adjusting the op amp stage gain. As the control voltage at R8 is readjusted, the OC1 photoresistances track these changes, automatically readjusting the op amp in accordance with the new control voltage setting.

75-Ω VIDEO PULSE AMPLIFIER

HA-5190 can drive the 75-Ω coaxial cable with signals up to 2.5 V pk-pk without the need for current boosting. In this circuit, the overall gain is approximately unity because of the impedance matching network.

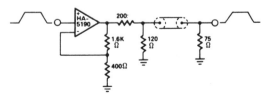

HARRIS   *Fig. 10-8*

VIDEO LINE DRIVING AMPLIFIER

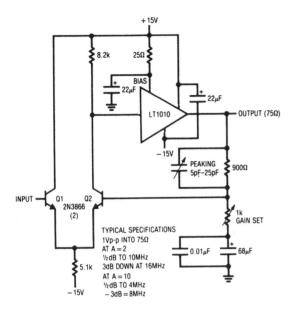

TYPICAL SPECIFICATIONS
1Vp-p INTO 75Ω
AT A = 2
½dB TO 10MHz
3dB DOWN AT 16MHz
AT A = 10
½dB TO 4MHz
−3dB = 8MHz

Q1 and Q2 form a differential stage which single-ends into the LT1010. The capacitively terminated feedback divider gives the circuit a dc gain of 1, while allowing ac gains up to 10. Using a 20-Ω bias resistor, the circuit delivers 1 V pk-pk into a typical 75-Ω video load. For applications sensitive to NTSC requirements, dropping the bias resistor value will aid performance. At $A = 2$, the gain is within 0.5 dB to 10 MHz and the −3 dB point occurs at 16 MHz. At $A = 10$, the gain is flat, within ±0.5 dB to 4 MHz, and the −3 dB point occurs at 8 MHz. The peaking adjustment should be optimized under loaded output conditions.

LINEAR TECHNOLOGY CORP.   *Fig. 10-9*

SUMMING AMPLIFIER/CLAMPING CIRCUIT

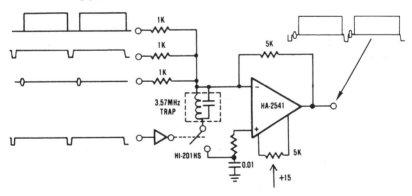

HARRIS   *Fig. 10-10*

This circuit is a traditional summing amplifier configuration with the addition of the dc clamping circuit. The operation is quite simple; each component—synchronization, color burst, picture information, etc.—of the composite video signal is applied to its own input terminal of the amplifier. These signals combine algebraically and form the composite signal at the output. The clamping circuit, if used, restores the 0-V reference of the composite signal.

TWO-STAGE WIDEBAND AMPLIFIER

This wideband high-gain configuration uses two SL550s connected in series. The first stage is connected in common-emitter configuration, the second stage is a common-base circuit. Stable gains of up to 65 dB can be achieved by the proper choice of R1 and R2. The bandwidth is 5 to 130 MHz, with a noise figure only marginally greater than the 2.0 dB specified for a single-stage circuit.

PLESSEY

Fig. 10-11

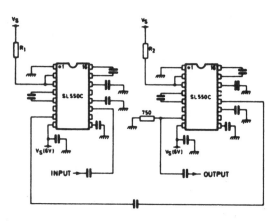

ALL CAPACITORS 1000 pF

VIDEO IF AMPLIFIER AND LOW-LEVEL VIDEO DETECTOR

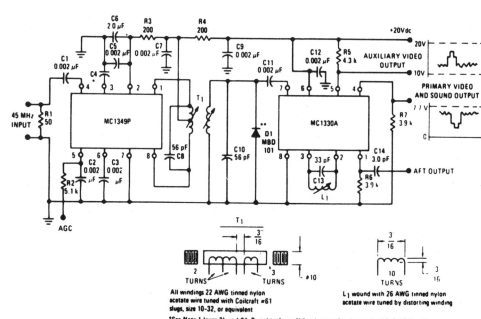

*See Note 1 (page 3), and C4, Parts List (page 4) for this specification on the MC1349P Data Sheet
**See Input Overload Section of the Design Characteristics Page 3, and General Information, Page 5, Note 6

MOTOROLA

Fig. 10-12

TELEVISION IF AMPLIFIER
AND DETECTOR

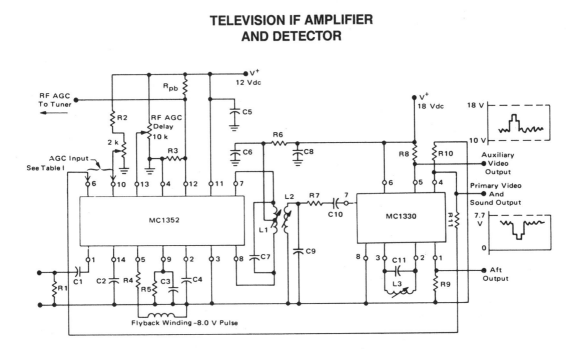

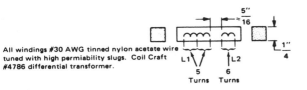

C10 = 62 pF
C11 = (See Table II)
All Resistors 1/4-Watt ±5%

All windings #30 AWG tinned nylon acetate wire tuned with high permiability slugs. Coil Craft #4786 differential transformer.

L1 5 Turns L2 6 Turns

See Table II

Wound with #26 AWG tinned nylon acetate wire tuned by distorting winding.

TABLE I

Video Polarity	Pin 6 Voltage	Pin 10 Voltage	R4
Negative-Going Sync.	5.5 / 2.0 / 0	Adj. 1.0–4.0 Vdc / Nom 2.0 V	0
Positive-Going Sync.	Adj. 1.0–8.0 Vdc / Nom 4.5 V	4.5 / 0	3.9 k

TABLE II

Component	36 MHz	45 MHz	58 MHz
C7	24 pF	15 pF	10 pF
C9	18 pF	12 pF	10 pF
C11	33 pF	33 pF	18 pF
L3	12 Turns	10 Turns	10 Turns

R_{pb} (See Text)
R1 = 50 Ω
R2 = 3.9 kΩ
R3 = (See Text)
R4 = (See Table I)
R5 = 220 kΩ
R6 = 220 Ω
R7 = 22 Ω
R8 = 3.3 kΩ
R9 = 3.9 kΩ

R10 = 3.9 kΩ
R11 = 4.7 kΩ
C1 = 0.001 μF
C2 = 0.1 μF
C3 = 0.25 μF ·
C5 = 0.1 μF
C6 = 0.1 μF
C7 = (See Table II)
C8 = 0.1 μF
C9 = (See Table II)

MOTOROLA

Fig. 10-13

TV SOUND IF OR FM IF AMPLIFIER WITH QUADRATURE DETECTOR

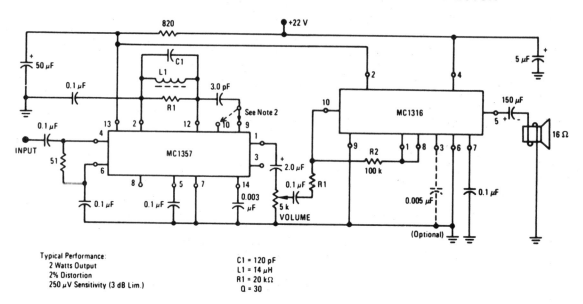

Typical Performance:
2 Watts Output
2% Distortion
250 μV Sensitivity (3 dB Lim.)

C1 = 120 pF
L1 = 14 μH
R1 = 20 kΩ
Q = 30

MOTOROLA

Fig. 10-14

VOLTAGE-CONTROLLED VARIABLE-GAIN AMPLIFIER

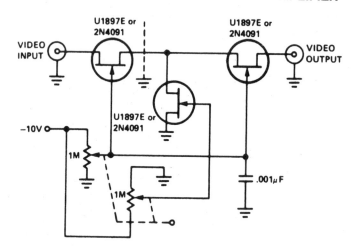

TELEDYNE

Fig. 10-15

The tee attenuator provides for optimum dynamic linear-range attenuation up to 100 dB, even at $f =$ 10.7 MHz, with proper layout.

JFET-BIPOLAR CASCODE VIDEO AMPLIFIER

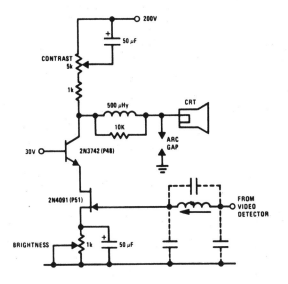

The JFET-bipolar cascode circuit will provide full video output for the CRT cathode drive. Gain is about 90. The cascode configuration eliminates Miller-capacitance problems with the 2N4091 JFET, allowing direct drive from the video detector. An m-derived filter using stray capacitance and a variable inductor prevents the 4.5 MHz sound frequency from being amplified by the video amplifier.

NATIONAL SEMICONDUCTOR

Fig. 10-16

VIDEO AMPLIFIER

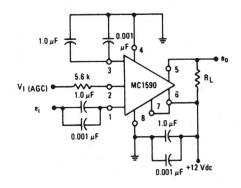

MOTOROLA

Fig. 10-17

VIDEO AMPLIFIER

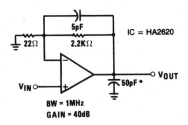

IC = HA2620

BW = 1MHz
GAIN = 40dB

*A small load capacitance of at least 30pF (including stray capacitance) is recommended to prevent possible high frequency oscillations.

HARRIS

Fig. 10-18

Sources

Chapter 1

Fig. 1-1. Hands-On Electronics, 7-8/86, p. 43.

Fig. 1-2. Reprinted with the permission of National Semiconductor Corp., Audio/Radio Handbook, 1980, p. 4-44.

Fig. 1-3. Reprinted with the permission of National Semiconductor Corp., Audio/Radio Handbook, 1980, p. 4-44.

Fig. 1-4. GE/RCA, BiMOS Operational Amplifiers Circuit Ideas, 1987, p. 25.

Fig. 1-5. Popular Electronics.

Fig. 1-6. Reprinted with the permission of National Semiconductor Corp., Audio/Radio Handbook, 1980, p. 4-20.

Fig. 1-7. Radio-Electronics, 3/86, p. 59.

Fig. 1-8. QST, 1/89, p. 20.

Fig. 1-9. Hands-On Electronics/Popular Electronics, 11/88, p. 39.

Fig. 1-10. Hands-On Electronics, Spring 1985, p. 36.

Fig. 1-11. Reprinted with the permission of National Semiconductor Corp., Linear Databook, 1982, p. 10-171.

Fig. 1-12. Reprinted with the permission of National Semiconductor Corp., Linear Databook, 1982, p. 10-63.

Fig. 1-13. Texas Instruments, Linear and Interface Circuits Applications, 1985, Vol. 1, p. 3-14.

Fig. 1-14. Signetics, 1987 Linear Data Manual Vol. 1: Communications, 11/6/86, p. 7-246.

Fig. 1-15. Reprinted with the permission of National Semiconductor Corp., Audio/Radio Handbook, 1980, p. 4-14.

Fig. 1-16. Reprinted with the permission of National Semiconductor Corp., Transistor Databook, 1982, p. 7-23.

Fig. 1-17. Signetics, 1987 Linear Data Manual Vol. 1: Communications, 11/86, p. 7-251.

Fig. 1-18. National Semiconductor Corp., Linear Applications Databook, p. 1065.

Fig. 1-19. Reprinted with the permission of National Semiconductor Corp., Audio/Radio Handbook, 1980, p. 4-51.

Fig. 1-20. Reprinted with permission of National Semiconductor Corp., Audio/Radio Handbook, 1980, p. 4-51.

Fig. 1-21. Courtesy of Fairchild Camera and Instrument

Corp., Linear Databook, 1982, p. 4-89.

Fig. 1-22. Reprinted with the permission of National Semiconductor Corp., Linear Databook, 1980, p. 10-203.

Fig. 1-23. Courtesy of Fairchild Camera and Instrument Corp., Fairchild Progress, 5-6/77, p. 22.

Fig. 1-24. Siliconix, MOSpower Applications Handbook, p. 6-101.

Fig. 1-25. Motorola, Motorola TMOS Power FET Design Ideas, 1985, p. 1.

Fig. 1-26. Courtesy of Fairchild Camera and Instrument Corp., Fairchild Progress, 5-6/77, p. 22.

Fig. 1-27. National Semiconductor Corp., Linear Databook, 1982, p. 3-187.

Fig. 1-28. Signetics, 1987 Linear Data Manual, Vol. 2: Industrial, 11/86, p. 4-135.

Fig. 1-29. Motorola, Motorola Power FET Design Ideas, 1985, p. 2.

Fig. 1-30. Hands-On Electronics, 5/87, p. 96.

Fig. 1-31. Harris, Analog Product Data Book, 1988, p. 10-161.

Fig. 1-32. Siliconix, MOSpower Applications Handbook, p. 6-180.

Fig. 1-33. Radio-Electronics, 8/88, p. 33.

Fig. 1-34. Signetics, RF Communications Handbook, 1989, p. 1-61 and 1-62.

Fig. 1-35. Reprinted with the permission of National Semiconductor Corp., Hybrid Products Databook, 1982, p. 17-170.

Fig. 1-36. Reprinted with the permission of National Semiconductor Corp., Application Note AN69, p. 4.

Fig. 1-37. Popular Electronics, 7/89, p. 26.

Fig. 1-38. Signetics, Signetics Analog Data Manual, 1983, p. 10-99.

Fig. 1-39. Hands-On Electronics, 5/87, p. 96.

Chapter 2

Fig. 2-1. Signetics, Signetics Analog Data Manual, 1982, p. 3-90.

Fig. 2-2. GE/RCA, BiMOS Operational Amplifiers Circuit Ideas, 1987, p. 21.

Fig. 2-3. Texas Instruments, Linear and Interface Circuits Applications, Vol. 1, 1985, p. 3-17.

Fig. 2-4. Radio-Electronics, 7/70, p. 38.

Fig. 2-5. Hands-On Electronics, 7-8/86, p. 16.

Fig. 2-6. Reprinted with the permission of National Semiconductor Corp., Audio/Radio Handbook, 1980, p. 2-45.

Fig. 2-7. Reprinted with the permission of National Semiconductor Corp., Audio/Radio Handbook, 1980, p. 2-43.

Fig. 2-8. 73 Amateur Radio, 12/76, p. 170.

Fig. 2-9. GE/RCA, BiMOS Operational Amplifiers Circuit Ideas, 1987, p.21.

Fig. 2-10. Canadian Projects No. 1, Spring 78, p. 27.

Fig. 2-11. Reprinted with the permission of National Semiconductor Corp., Application Note AN69, p. 4.

Fig. 2-12. Reprinted with the permission of National Semiconductor Corp., Linear Databook, 1982, p. 10-25.

Fig. 2-13. Courtesy of Motorola Inc., Linear Integrated Circuits, 1979, p. 5-17.

Fig. 2-14. Reprinted with the permission of National Semiconductor Corp., Linear Databook, 1982, p. 10-170.

Fig. 2-15. Reprinted with the permission of National Semiconductor Corp., Transistor Databook, 1982, p. 11-29.

Fig. 2-16. Reprinted with the permission of National Semiconductor Corp., Data Conversion/Acquisition Databook, 1980, p. 3-91.

Fig. 2-17. Reprinted with the permission of National Semiconductor Corp., Audio/Radio Handbook, 1980, p. 2-21.

Fig. 2-18. Signetics, Signetics Analog Data Manual, 1977, p. 466.

Fig. 2-19. Signetics, Signetics Analog Data Manual, 1983, p. 10-92.

Fig. 2-20. Signetics, Signetics Analog Data Manual, 1982, p. 15-6.

Fig. 2-21. Courtesy of Motorola Inc., Motorola Semiconductor Library, Vol. 6, Series B, p. 8-21.

Fig. 2-22. Courtesy of Motorola Inc., Motorola Semiconductor Library, Vol. 6, Series B, p. 8-21.

Fig. 2-23. Courtesy of Motorola Inc., Motorola Semiconductor Library, Vol. 6, Series B, p. 8-21.

Fig. 2-24. Reprinted with the permission of National Semiconductor Corp., National Semiconductor Application Note AN125, p. 7.

Fig. 2-25. Signetics Analog Data Manual, 1982, p. 15-6.

Fig. 2-26. Signetics Analog Data Manual, 1977, p. 466.

Fig. 2-27. Reprinted with the permission of National Semiconductor Corp., Data Conversion/Acquisition Databook, 1980, p. 3-88.

Fig. 2-28. Reprinted with the permission of National Semiconductor Corp., Audio/Radio Handbook, 1980, p. 2-20.

Fig. 2-29. Signetics Analog Data Manual, 1982, 3-90.

Fig. 2-30. Harris, Analog Product Data Book, 1988, p. 10-108.

Fig. 2-31. Hands-On Electronics, Summer 1984, p. 74.

Fig. 2-32. Signetics Analog Data Manual, 1983, p. 10-20.

Fig. 2-33. ZeTeX (formerly Ferranti), Technical Handbook Super E-Line Transistors, 1987, p. SE-153.

Fig. 2-34. Motorola Inc., Linear Integrated Circuits, 1979, p. 6-58.

Fig. 2-35. Motorola Inc., MR-E Experimenters Handbook, p. 158.

Fig. 2-36. Canadian Projects, No. 1, Spring 1978, p. 29.

Fig. 2-37. Popular Electronics, Fact Card No. 110.

Fig. 2-38. Hands-On Electronics, Summer 1984, p. 74.

Chapter 3

Fig. 3-1. Linear Technology, Application Note 9, p. 6.

Fig. 3-2. © Siliconix Inc., Application Note, AN73-6, p. 3.

Fig. 3-3. Reprinted with the permission of National Semiconductor Corp., Hybrid Products Databook, 1982, p. 3-7.

Fig. 3-4. Maxim, Maxim Advantage, p. 45.

Fig. 3-5. Linear Technology Corp., Linear Applications Handbook, 1987, p. AN3-2.

Fig. 3-6. GE/RCA, BiMOS Operational Amplifiers Circuit Ideas, 1987, p. 19.

Fig. 3-7. Courtesy of Motorola Inc., Linear Integrated Circuits, 1979, p. 3-31.

Fig. 3-8. Siliconix, Siliconix Analog Switch and IC Product Data Book, 1/83, p. 6-21.

Fig. 3-9. Siliconix, Siliconix Analog Switch and IC Product Data Book, 1/83, p. 6-15.

Fig. 3-10. Precision Monolithics Inc., 1981 Full Line Catalog, p. 16-37.

Fig. 3-11. Harris Semiconductor, Linear and Data Acquisition Products, 1977, p. 2-85.

Fig. 3-12. Electronic Engineering, 9/84, p. 33.

Fig. 3-13. Texas Instruments, Linear and Interface Circuits Applications Vol. 1, 1985, p. 3-18.

Fig. 3-14. Linear Technology Corp., Linear Databook, 1986, p. 2-83.

Fig. 3-15. Hands-On Electronics, 12/86, p. 42.

Fig. 3-16. Linear Technology Corp., Linear Databook, 1986, p. 2-82.

Fig. 3-17. Reprinted with the permission of National Semiconductor Corp., National Semiconductor Application Note LB1, p. 2.

Fig. 3-18. Courtesy of Texas Instruments Inc., Linear Control Circuits Data Book, Second Edition, p. 120.

Fig. 3-19. Harris, Analog Product Data Book, 1988, p. 10-181.

Fig. 3-20. Maxim, Maxim Advantage, p. 45.

Fig. 3-21. GE/RCA, BiMOS Operational Amplifiers Circuit Ideas, 1987, p. 15.

Fig. 3-22. © Siliconix Inc., T100/T300 Applications.

Fig. 3-23. Reprinted with the permission of National Semiconductor Corp., Data Conversion/Acquisition Databook, 1980, p. 4-27.

Fig. 3-24. Reprinted with the permission of National Semiconductor Corp., Data Conversion/Acquisition Databook, 1980, p. 3-102.

Fig. 3-25. Courtesy of Motorola Inc., Linear Integrated Circuits, 1979, p. 3-82.

Fig. 3-26. Reprinted with permission of Analog Devices Inc., Data Acquisition Databook, 1982, p. 4-92.

Fig. 3-27. Precision Monolithics Inc., 1981 Full Line catalog, p. 6-50.

Fig. 3-28. Precision Monolithics Inc., 1981 Full Line catalog, p. 16-37.

Fig. 3-29. Signetics Analog Data Manual, 1982, 3-15.

Fig. 3-30. Reprinted with the permission of National Semiconductor Corp., Linear Applications Handbook, 1982, p. AN242-15.

Fig. 3-31. Signetics, Analog Data Manual, 1982, p. 3-71.

Fig. 3-32. Precision Monolithics Inc., 1981 Full Line Catalog, p. 6-77.

Fig. 3-33. Courtesy of Fairchild Camera and Instrument Corp., Linear Databook, 1982, p. 4-178.

Fig. 3-34. Linear Technology Corp., Linear Applications Handbook, 1987, p. AN18-3

Fig. 3-35. Harris, Analog Product Data Book, 1988, p. 10-150.

Fig. 3-36. Electronic Engineering, 9/78, p. 17.

Fig. 3-37. Electronic Engineering, 9/84, p. 33.

Fig. 3-38. Courtesy of Fairchild Camera and Instrument Corp., Linear Databook, 1982, p. 4-43.

Fig. 3-39. Reprinted with the permission of National Semiconductor Corp., National Semiconductor Application Note 32, p. 5.

Fig. 3-40. Precision Monolithics Inc., 1981 Full Line Catalog, p. 6-171.

Fig. 3-41. Courtesy of Texas Instruments Inc., Linear Control Circuits Data Book, Second Edition, p. 122.

Fig. 3-42. Precision Monolithics Inc., 1981 Full Line Catalog, p. 7-11.

Fig. 3-43. Precision Monolithics Inc., 1981 Full Line Catalog, p. 7-6.

Fig. 3-44. Precision Monolithics Inc., 1981 Full Line Catalog, p. 16-159.

Fig. 3-45. Reprinted with permission of Analog Devices Inc., Data Acquisition Databook, 1982, p. 4-56.

Fig. 3-46. Harris, Analog Product Data Book, 1988, p. 10-110.

Fig. 3-47. Linear Technology Corp., Linear Applications Handbook, 1987, p. AN21-5.

Fig. 3-48. Texas instruments, Linear and Interface Circuits Applications, Vol. 1, 1985, p. 3-2, 3-4.

Fig. 3-49. Harris, Analog Product Data Book, 1988, p. 10-174.

Fig. 3-50. Signetics, 1987 Linear Data Manual Vol. 1: Communications, 3/87, p. 4-345.

Fig. 3-51. Courtesy of Fairchild Camera and Instrument Corp., Linear Databook, 1982, p. 9-17.

Fig. 3-52. NASA, NASA Tech Briefs, Spring 1983, p. 244.

Fig. 3-53. Siliconix, Integrated Circuits Data Book, 1988, p. 5-172.

Chapter 4

Fig. 4-1. Siliconix, Integrated Circuits Data Book, 3/85, p. 10-85.

Fig. 4-2. Siliconix, Integrated Circuits Data Book, 3/85, p. 10-79.

Fig. 4-3. Siliconix, Integrated Circuits Data Book, 3/85, p. 2-144.

Fig. 4-4. Signetics, 1987 Linear Data Manual Vol. 2: Industrial, 10/86, p. 4-261.

Fig. 4-5. Siliconix, Integrated Circuits Data Book, 3/85, p. 2-103.

Chapter 5

Fig. 5-1. Hands-On Electronics, Summer 1984, p. 77.

Fig. 5-2. Courtesy William Sheets.

Fig. 5-3. Courtesy William Sheets.

Fig. 5-4. Electronic Engineering, 5/84, p. 44.

Fig. 5-5. Texas Instruments, Linear and Interface Circuits Applications, Vol. 1, 1985, p. 3-13.

Fig. 5-6. Courtesy of Motorola Inc., Linear Integrated Circuits, 1979, p. 6-23.

Fig. 5-7. Texas Instruments, Linear and Interface Circuits Data Book, Second Edition, p. 145.

Fig. 5-8. Electronics Today International, 4/85, p. 82.

Chapter 6

Fig. 6-1. Siliconix, Integrated Circuits Data Book, 3/85, p. 10-211.

Fig. 6-2. Intersil, Applications Handbook, 1988, p. 3-181.

Fig. 6-3. Electronic Design 15, 7/75, p. 68.

Fig. 6-4. Reprinted with permission of Analog Devices Inc., Data Acquisition Databook, 1982, p. 4-98.

Fig. 6-5. Reprinted with the permission of National Semiconductor Corp., National Semiconductor Application Note AN125, p. 3.

Fig. 6-6. Precision Monolithics Inc., 1981 Full Line Catalog, p. 16-160.

Fig. 6-7. Signetics Analog Data Manual, 1982, p. 3-103.

Fig. 6-8. Precision Monolithics Inc., 1981 Full Line Catalog, p. 6-127.

Fig. 6-9. Courtesy of Motorola Inc., Linear Integrated Circuits, 1979, p. 3-83.

Fig. 6-10. Electronics Today International, 2/82, p. 58.

Fig. 6-11. Signetics, Analog Data Manual , 1983, p. 10-100.

Fig. 6-12. Maxim, Maxim Advantage, p. 44.

Fig. 6-13. Electronic Engineering, 9/88, p. 28.

Fig. 6-14. Siliconix, Integrated Circuits Data Book, 1988, p. 13-166.

Fig. 6-15. GE/RCA, BiMOS Operational Amplifiers Circuit Ideas, 1987, p.13.

Fig. 6-16. Hands-On Electronics, Fact Card No. 29.

Fig. 6-17. Hands-On Electronics, Fact Card No. 29.

Fig. 6-18. GE/RCA, BiMOS Operational Amplifiers Circuit Ideas, 1987, p.13.

Fig. 6-19. Harris, Analog Product Data Book, 1988, p. 10-95.

Fig. 6-20. Harris, Analog Product Data Book, 1988, p. 10-96.

Fig. 6-21. NASA, NASA Tech Briefs, 9-10/86, p. 43.

Fig. 6-22. Electronic Engineering, 11/85, p. 32.

Fig. 6-23. Signetics, 1987 Linear Data Manual Vol 1: Communications, 3/87, p. 4-345.

Fig. 6-24. Linear Technology Corp., Linear Applications Handbook 1987, p. AN21-1.

Fig. 6-25. Signetics, 1987 Linear Data Manual, Vol. 2: Industrial, 10/86, p. 4-260.

Fig. 6-26. Linear Technology Corp., Linear Applications Handbook 1987, p. AN21-2.

Fig. 6-27. Courtesy of Motorola Inc., Linear Integrated Circuits, 1979, p. 3-83.

Fig. 6-28. Courtesy of Fairchild Camera and Instrument Corp., Linear Databook, 1982, p. 4-41.

Fig. 6-29. Siliconix, Analog Switch and IC Product Data Book, 1/82, p. 7-56.

Fig. 6-30. Reprinted with permission of Analog Devices Inc., Data Acquisition Databook, 1982, p. 4-119.

Fig. 6-31. Courtesy of Fairchild Camera and Instrument Corp., Linear Databook, 1982, p. 4-42.

Fig. 6-32. Courtesy of Motorola Inc., *Linear Integrated Circuits*, 1979, p. 3-17.

Fig. 6-33. Courtesy of Motorola Inc., *Linear Integrated Circuits*, 1979, p. 3-131.

Fig. 6-34. Harris, *Analog Data Book*, 1984.

Fig. 6-35. Intersil, *Intersil Data Book*, 5/83, p. 5-36.

Fig. 6-36. Precision Monolithics Inc., *1981 Full Line Catalog*, p. 16-37.

Fig. 6-37. Signetics, *Analog Data Manual*, 1983, p. 17-17.

Fig. 6-38. Intersil, *Intersil Data Book*, 5/83, p. 5-36.

Fig. 6-39. Courtesy of Motorola Inc., *Linear Integrated Circuits*, 1979, p. 3-17.

Fig. 6-40. Reprinted with the permission of National Semiconductor Corp., *Hybrid Products Databook*, 1982, p. 1-83.

Fig. 6-41. Reprinted with the permission of National Semiconductor Corp., *Data Conversion/Acquisition Databook*, 1980, p. 3-107.

Fig. 6-42. Reprinted with the permission of National Semiconductor Corp., *Transistor Databook*, 1982, p. 11-29.

Fig. 6-43. Signetics, *Analog Data Manual*, 1977, p. 35.

Fig. 6-44. Courtesy of Fairchild Camera and Instrument Corp., *Linear Databook*, 1982, p. 5-39.

Fig. 6-45. Precision Monolithics Inc., *1981 Full Line Catalog*, p. 6-10.

Fig. 6-46. Courtesy of Motorola Inc., *Motorola Semiconductor Library*, Vol. 6, Series B, p. 8-21.

Fig. 6-47. Courtesy of Fairchild Camera and Instrument Corp., *Linear Databook*, 1982, p. 4-119.

Fig. 6-48. Signetics, *Analog Data Manual*, 1982, p. 3-83.

Fig. 6-49. *Electronic Engineering*, 6/83, p. 31.

Fig. 6-50. Reprinted with the permission of National Semiconductor Corp., *Audio/Radio Handbook*, 1980, p. 2-67.

Fig. 6-51. Reprinted with the permission of National Semiconductor Corp., *Hybrid Products Databook*, 1982, p. 77.

Fig. 6-52. *Electronic Engineering*, Applied Ideas, 9/88, p. 25.

Fig. 6-53. Maxim, *Maxim Advantage*, p. 44.

Fig. 6-54. Harris, *Analog Product Data Book*, 1988, p. 10-167.

Fig. 6-55. Reprinted from *EDN*, 1/79, © 1989 Cahners Publishing Co., a division of Reed Publishing USA.

Fig. 6-56. GE/RCA, *BiMOS Operational Amplifiers Circuit Ideas*, 1987, p. 20.

Fig. 6-57. Signetics, *1987 Linear Data Manual*, Vol. 1: Communications, 8/87, p. 4-346.

Fig. 6-58. Signetics, *1987 Linear Data Manual* Vol. 2: Industrial, 11/6/86, p. 4-136.

Fig. 6-59. Signetics, *1987 Linear Data Manual* Vol. 2: Industrial, 2/87, p. 4-243.

Fig. 6-60. *Ham Radio*, 9/84, p. 24.

Fig. 6-61. Linear Technology Corp., *Linear Databook Supplement*, 1988, p. S2-34.

Fig. 6-62. Siliconix, *Integrated Circuits Data Book*, 3/85, p. 10-154.

Fig. 6-63. National Semiconductor Corp., *Transistor Databook*, 1982, p. 11-23.

Fig. 6-64. Linear Technology Corp., *Linear Applications Handbook 1987*, p. AN21-1.

Fig. 6-65. Signetics, *1987 Linear Data Manual* Vol. 2: Industrial, 10/86, p. 4-260.

Fig. 6-66. Reprinted with permission of National Semiconductor Corp., *Application Note AN125*, p. 2.

Chapter 7

Fig. 7-1. Harris, *Analog Product Data Book*, 1988, p. 10-13.

Fig. 7-2. Harris, *Analog Product Data Book*, 1988, p. 10-13.

Fig. 7-3. Linear Technology Corp., *Linear Applications Handbook 1987*, p. AN35.

Fig. 7-4. Siliconix, *Integrated Circuits Data Book*, 1988, p. 5-128.

Fig. 7-5. Intersil, *Applications Handbook*, 1988, p. 2-34.

Fig. 7-6. Precision Monolithics Inc., *1981 Full Line Catalog*, p. 12-50.

Fig. 7-7. Harris, *Analog Product Data Book*, 1988, p. 10-169.

Fig. 7-8. National Semiconductor Corp., *Transistor Databook*, 1982, p. 11-25.

Chapter 8

Fig. 8-1. Motorola, *RF Data Manual*, 1986, p. 6-182.

Fig. 8-2. *Radio-Electronics*, 3/82, p. 59.

Fig. 8-3. Motorola, *RF Data Manual*, 1986, p. 6-141.

Fig. 8-4. *QST*, 7/87, p. 31.

Fig. 8-5. *Electronics Today International*, Summer 1982, p. 45.

Fig. 8-6. *73 Amateur Radio*, p. 31.

Fig. 8-7. Motorola, *RF Data Manual*, 1986, p. 6-232.

Fig. 8-8. Motorola, *RF Data Manual*, 1986, p. 6-239.

Fig. 8-9. Courtesy of Motorola Inc., *Communications Engineering Bulletin*, EB-46.

Fig. 8-10. Motorola, *RF Data Manual*, 1986, p. 6-85.

Fig. 8-11. Courtesy of Motorola Inc., Communications Engineering Bulletin, EB-67.

Fig. 8-12. Motorola, RF Data Manual, 1986, p. 6-181.

Fig. 8-13. Motorola, RF Data Manual, 1986, p. 6-226.

Fig. 8-14. Courtesy of Motorola Inc., Application Note AN593, p. 6.

Fig. 8-15. Microwaves and RF, 1/83, p. 89.

Fig. 8-16. Ham Radio, 7/86, p. 50.

Fig. 8-17. Motorola, RF Data Manual, 1986, p. 6-221.

Fig. 8-18. Courtesy of Motorola Inc., Motorola Semiconductor Library, Vol. 6, Series B, p. 8-58.

Fig. 8-19. Courtesy of Motorola Inc., Motorola Semiconductor Library, Vol. 6, Series B, p. 8-58.

Fig. 8-20. © Siliconix Inc., MOSpower Design Catalog, 1/83, p. 5-10.

Fig. 8-21. Teledyne Semiconductor, Data and Design Manual, 1981, p. 11-178.

Fig. 8-22. Teledyne Semiconductor, Data and Design Manual, 1981, p. 11-178.

Fig. 8-23. Teledyne Semiconductor, Data and Design Manual, 1981, p. 11-178.

Fig. 8-24. QST, 9/85, p. 41.

Fig. 8-25. Microwaves and RF, 9/85, p. 191.

Fig. 8-26. Teledyne Semiconductor, Data and Design Manual, 1981, p. 11-178.

Fig. 8-27. Reprinted with the permission of National Semiconductor Corp., Transistor Databook, 1982, p. 11-33.

Fig. 8-28. © Siliconix Inc., MOSpower Design Catalog, 1-83, p. 5-10.

Fig. 8-29. Motorola, RF Data Manual, 1986, p. 6-141.

Fig. 8-30. Teledyne Semiconductor, Data and Design Manual, 1981, p. 11-178.

Fig. 8-31. Signetics, Analog Data Manual, 1983, p. 17-18.

Fig. 8-32. Signetics, Analog Data Manual, 1983, p. 17-15.

Fig. 8-33. 73 Amateur Radio.

Fig. 8-34. QST, 5/86, p. 23.

Fig. 8-35. Signetics, RF Communications Handbook, 1989, p. 1-31.

Fig. 8-36. NASA, NASA Tech Briefs, Spring 1984, p. 322.

Fig. 8-37. Courtesy of Motorola Inc., Motorola Semiconductor Library, Vol. 6, Series B, p. 8-59.

Fig. 8-38. © Siliconix Incorporated. MOSPOWER Design Catalog, 1/83, p. 5-36.

Fig. 5-39. 73 Amateur Radio, 3/89, p. 66.

Fig. 8-40. Harris, Analog Product Data Book, 1988, p. 10-58.

Fig. 8-41. GE/RCA, BiMOS Operational Amplifiers Circuit Ideas, 1987, p. 20.

Fig. 8-42. Siliconix, Small-Signal FET Data Book, 1989, p. 4-158, 4-159, and 9-42.

Fig. 8-43. R-E Experimenters Handbook, 1989, p. 146.

Fig. 8-44. R-E Experimenters Handbook, 1989, p. 33.

Fig. 8-45. Reprinted from EDN, 1/7/88, © 1989 Cahners Publishing Co., a division of Reed Publishing USA.

Fig. 8-46. Reprinted from EDN, 5/5/78, © 1989 Cahners Publishing Co., a division of Reed Publishing USA.

Fig. 8-47. Hands-On Electronics, 7-8/86.

Fig. 8-48. Maxim, 1986 Power Supply Circuits, p. 27.

Fig. 8-49. Texas Instruments, Linear and Interface Circuits Applications, 1985, Vol. 1, p. 6-36.

Fig. 8-50. Texas Instruments, Linear and Interface Circuits Applications, 1985, Vol. 1, p. 6-36.

Chapter 9

Fig. 9-1. GE/RCA, BiMOS Operational Amplifiers Circuit Ideas, 1987, p. 18.

Fig. 9-2. GE/RCA, BiMOS Operational Amplifiers Circuit Ideas, 1987, p. 21.

Fig. 9-3. GE/RCA, BiMOS Operational Amplifiers Circuit Ideas, 1987, p. 19.

Fig. 9-4. Signetics, 1987 Linear Data Manual Vol. 1: Communications, 2/87, p. 7-173.

Fig. 9-5. Signetics, 1987 Linear Data Manual Vol. 1: Communications, 2/87, p. 7-173.

Fig. 9-6. Radio-Electronics, 3/86, p. 8.

Fig. 9-7. Signetics, 1987 Linear Data Manual Vol. 1: Communications, 2/87, p. 7-173.

Fig. 9-8. Signetics, 1987 Linear Data Manual Vol. 1: Communications, 2/87, p. 7-173.

Fig. 9-9. Reprinted from EDN, 9/9/88, © 1989 Cahners Publishing Co., a division of Reed Publishing USA.

Fig. 9-10. Reprinted with permission of National Semiconductor Corp., Transistor Databook, 1982, p. 11-28.

Fig. 9-11. Signetics, Signetics Analog Data Manual, 1982, p. 4-8.

Fig. 9-12. Signetics, Signetics Analog Data Manual, 1982, p. 15-6.

Fig. 9-13. Signetics, Signetics Analog Data Manual, 1977, p. 466.

Fig. 9-14. Reprinted with permission of National Semiconductor Corp., Audio/Radio Handbook, 1980, p. 2-27.

Fig. 9-15. Reprinted with permission of National Semiconductor Corp., Audio/Radio Handbook, 1980, p. 2-32.

Fig. 9-16. Electronics Today International, 4/79, p. 18.

Fig. 9-17. Reprinted with permission of National Semiconductor Corp., Linear Databook, 1989, p. 3-389.

Chapter 10

Fig. 10-1. GE, Application Note 90.88, p. 7.

Fig. 10-2. Courtesy of Motorola Inc., Application Note AN-545A, p. 12.

Fig. 10-3. Harris, Analog Product Data Book, 1988, p. 10-96.

Fig. 10-4. Reprinted with permission from Electronic Design. © 1989 Penton Publishing.

Fig. 10-5. Reprinted with the permission of National Semiconductor Corp., Transistor Databook, 1982, p. 7-26.

Fig. 10-6. Reprinted with the permission of National Semiconductor Corp., Transistor Databook, 1982, p. 11-31.

Fig. 10-7. Harris, Analog Product Data Book, 1988, p. 10-58.

Fig. 10-8. Harris, Analog Product Data Book, 1988, p. 10-54.

Fig. 10-9. Linear Technology Corp., Linear Applications Handbook, 1987, p. AN4-3.

Fig. 10-10. Harris, Analog Product Data Book, 1988, p. 10-149.

Fig. 10-11. Plessey Semiconductors, Linear IC Handbook, 5/82, p. 129.

Fig. 10-12. Courtesy of Motorola Inc., Linear Integrated Circuits, 1979, p. 5-50.

Fig. 10-13. Courtesy of Motorola Inc., Application Note AN-545A, p. 12.

Fig. 10-14. Courtesy of Motorola Inc., Linear Integrated Circuits, 1979, p. 5-73.

Fig. 10-15. Teledyne Semiconductor, Data and Design Manual, 1981, p. 11-207.

Fig. 10-16. Reprinted with the permission of National Semiconductor Corp., Transistor Databook 1982, p. 11-30.

Fig. 10-17. Courtesy of Motorola, Inc., Motorola Semiconductor Library, Volume 6, Series B.

Fig. 10-18. Harris Semiconductor, Linear and Acquisition Products, 1977, p. 2-46.

Index